URBAN DEVELOPMENT AND NEW TOWNS IN THE THIRD WORLD

Dedicated to the poor of Navi Mumbai

Urban Development and New Towns in the Third World

Lessons from the New Bombay experience

ALAIN R.A. JACQUEMIN
School of Oriental and African Studies (SOAS)
and University of Brussels (VUB),
Currently Monitoring Expert for UNDP's
South Asia Poverty Alleviation Programme, Kathmandu

Routledge
Taylor & Francis Group

LONDON AND NEW YORK

First published 1999 by Ashgate Publishing

Reissued 2018 by Routledge
2 Park Square, Milton Park, Abingdon, Oxon, OX14 4RN
711 Third Avenue, New York, NY I 0017, USA

Routledge is an imprint of the Taylor & Francis Group, an informa business

Publisher's Note
The publisher has gone to great lengths to ensure the quality of this reprint but points out that some imperfections in the original copies may be apparent.

Disclaimer
The publisher has made every effort to trace copyright holders and welcomes correspondence from those they have been unable to contact.

A Library of Congress record exists under LC control number: 99072981

ISBN 13: 978-1-138-35935-2 (hbk)
ISBN 13: 978-1-138-35937-6 (pbk)
ISBN 13: 978-0-429-43377-1 (ebk)

Contents

Part I
Urban Development and Urban Problems
in the Third World

2 Urbanisation, Urban Growth and Planning in the Third World

3 Urban Development in India

4 Urban Development in Bombay

v

Part II
Birth of New Bombay and Implementation
of the New Bombay Project

5 Birth of New Bombay: Policies and Strategies

6 Growth and Development of New Bombay

7 Impact of Growth of New Bombay on Original Population

8 New Bombay, a City for All?

Part III
Considerations on Planning Concepts and Plan Implementation

9 Considerations on Planning Concepts and Implementation

10 General Conclusions

List of Figures, Tables and Photographs

List of figures

List of tables

Photographs

List of Abbreviations

APM	Agricultural Produce Markets
ARV	Annual Rateable Value
BEST	Bombay Electricity Supply and Transport Undertaking
BMC	Bombay Municipal Corporation
BMR	Bombay Metropolitan Region
BMRDA	Bombay Metropolitan Region Development Authority
BMRPB	Bombay Metropolitan Region Planning Board
BMTC	Bombay Metropolitan Transport Corporation
BPT	Bombay Port Trust
BSES	Bombay Suburban Electricity Supply
BUDP	Bombay Urban Development Project
CBOs	Community Based Organisations
CIDCO	City and Industrial Development Corporation of Maharashtra
DDA	Delhi Development Authority
DRS	Demand Registration Scheme
EDA	Early Development Area
EIC	East India Company
EWS	Economic Weaker Sections
FSI	Floor Space Index (or FAR)
GBMC	Greater Bombay Municipal Corporation
GBUA	Greater Bombay Urban Agglomeration
HIG	High Income Group
HUDCO	Housing and Urban Development Corporation
JNPT	Jawaharlal Nehru Port Trust
LIG	Low Income Group
MIDC	Maharashtra Industrial Development Corporation Ltd.
MSEB	Maharahstra State Electricity Board
MWSSB	Maharashtra Water and Sewerage Board
NBDP	New Bombay Development Plan
NMMC	Navi Mumbai Municipal Corporation
NMMT	Navi Mumbai Municipal Transport
NRIs	Non-Residential Indians
NTDA	New Town Development Authority
ONGC	Oil and Natural Gas Company
PAP(s)	Project Affected Persons (People)
PDS	Participatory Developers' Scheme

RBI	Reserve Bank of India
RP	Reserve Price
SES-95	Socio-Economic Survey 1995-96
SICOM	State Industrial and Investment Corp. of Maharashtra
ST	State Transport
TBIA	Thana-Belapur Industrial Area

Acknowledgements

A study of this kind can never be the product of pure personal effort, and I would therefore like to express my gratitude to all those who have helped me in one way or another.

I am extremely grateful to the Research Council of the University of Brussels (VUB), who generously financed a major part of the study, and to the university's Centre of Political Studies, who provided additional funding and all the necessary infrastructural support. I would also like to thank the Ministry of Education of the Flemish Community and the Indian Council for Cultural Relations (ICCR) for their contributions. Without their continuous financial support this study would not have been possible.

Practically and morally I am much indebted to a great number of people and institutions: In Brussels to the colleagues of the *Vakgroep Politieke Wetenschappen* for their good humour and their contribution to an enjoyable work environment; further, to Mia Lammens at the Department of Geography, who has helped me with the creation of several geographical maps, to Jo Buelens for his mapping contribution and numerous practical tips, and to Koen Vlassenroot for his excellent taxi services. In London I am very grateful to a number of friends who always arranged a bed for me when staying over. Waqar, Anja, Nadja, Alli and Ruf, Tinki and Judith, thank you all. In Bombay I was lucky to have The Salvation Army arranging accommodation for me on a long-term basis. Without this support in a time of great problems I would probably have left Bombay (and my research) at an early stage. Thanks also to the ICCR office in Bombay for all their practical help, and for the many cultural invitations that provided moments of peace and quiet in an otherwise hectic environment; to Udita Jhunjhunwala for the practical and moral support and for the invitations to some splendid marriages and the introduction to Bombay's night-life; and to the Belgian Consulate, for the homely atmosphere, the familiar newspapers, and the Belgian cheese, beers and chocolats at the gatherings.

In New Bombay I am much indebted to the City and Industrial Development Corporation of Maharashtra (CIDCO), who provided me with the necessary office, transportation and research facilities, and especially to its joint managing director G.S. Gill, its PR officer B.R. Gaikwad and its then manager town services R.N. Kasar, the latter who arranged excellent accommodation for me. Without their support my fieldwork would have been greatly hampered, if not impossible. The people who have helped me with my research in *CIDCO Bhavan* are numerous, and naming them all would be impossible. I will only mention those whose time and generosity I have exploited to the full, and who have always patiently tried to provide the data and information that I asked for. In this respect I particularly acknowledge the efforts of A. Bhattacharya, K.S. Bahuleyan, Ms. Raje, S. Satpathy, Ms. Bavadekar, M. Rameshkumar, Ms. Dikshit, S.S. Phadke, and Ms Uma Adusumilli.

Many thanks also to Isaac Prasadam for letting me occupy his best room for so long, and to the many friends at the YMCA Belapur, who confronted me too often with my poor chess skills, and to those at the ashram, who did not manage to fully convince me of the 'hidden powers' of Sahaja yoga. They have made my long stay in *Navi Mumbai* a most enjoyable experience.

I am also greatly indebted to all those people in high-level positions, in the government and outside, who generously spent their precious time in interviews: A.B. Sapre, S.B. Gogte, G.S. Gill, Shirish Patel, Charles Correa, J.B. D'Souza, D.M. Sukthankar, D.T. Joseph, P.C. Patil, Dr. Desai, A.R. Jadhav, M.G. Ahire, V.K. Phatak, V. Srinivasan, Suraj Kaeley, and Dr. Kashelikar. Equally, I would like to thank all those individuals and households whom I have interviewed *in the field*; in the villages, the nodes, the slums and the BUDP areas. In their own way, they may have taught me more of the processes of urban development in New Bombay than any book could do. Without their co-operation and dedication several parts of this thesis would not have been written.

A very distinct word of thanks must go to a number of people: to Ramachandran, who joined me on my field visits in New Bombay to do the translations. He not only is an excellent translator, but also a most pleasant person who became a friend on whom I could always count for numerous other things; to Ms. Uma in CIDCO, whose enthusiasm and determination were contagious. I will always remember our lively discussions and her continuous support; to Swapna Banerjee-Guha at the University of Bombay, who introduced me into the New Bombay reality academically, who provided me with some very valuable research directives, who continuously supported what I was doing, and who helped me in numerous other ways (not in the least in my monthly severe struggles with the university authorities to get hold of my grant).

I am, however, most indebted to Robert Bradnock, an excellent supervisor and most pleasant person who, despite his overloaded agenda as head of department at SOAS and many other obligations, cautiously guided me through the research process and gave me absolute freedom to do the kind of work I had planned to do. He also gave me the opportunity to gain practical fieldwork experience which, looking back, has been of great help in conducting my own research in New Bombay. Even at present, in my new function as the overall monitoring expert of the South Asia Poverty Alleviation Programme of UNDP, I am heavily relying on these early research experiences. For all that, many thanks Bob!

I am also very grateful to my lovely wife Régine, from whom, due to this research work, I have lived far too often separated. She has supported me throughout, and this book is therefore as much her effort as it is mine.

Finally, I would like to thank Oxford University Press Bombay for giving permission to use one of their maps, and Ashgate Publications for guiding me through the preparation of the manuscript and for the rapid publication schedule for the book.

The data that are included in the book have been verified several times, but it is nevertheless most likely that several errors and shortcomings will remain. For these I am alone responsible.

AJ,
Kathmandu,
1 June 1999

Preface

Since the 1950s, the world has seen a dramatic increase in urban population, with the number of *urbanites* increasing from about 300 million in 1950 (12.5% of the then total world population) to over 2.7 billion at present (45% of current world population). By 2015 the number of urban dwellers will have further increased to over 4 billion.

This dramatic increase in urban population has largely been the result of rapid urban growth in the Third World. In 1950, the number of people living in cities in the developing world was 317 million. Today, that number has increased to 1.7 billion, and by 2025, just a generation ahead, the total urban population of Asia, Africa and Latin America will be 4.1 billion, with 2.7 billion people living in Asian cities alone.

India and Bombay are some of the most dramatic examples of rapid urban growth in the developing world, with India currently having an urban population matching the total population of the United States. Bombay, or *Mumbai* as it was recently renamed, is India's largest and economically most important city. Whereas this city had a population of only 1.5 million in 1941, today that number has increased to over 15 million, with the population living in slums and squatter settlements usually estimated to be well over 50%. Although the growth pattern of Bombay may have been dramatic, it is by no means a unique case in India or South Asia. By 2015, not less than 8 of the 30 largest cities in the world will be located in South Asia, 5 of which in India alone.

One of the strategies frequently adopted in Third World countries to cope with the dramatic urban growth of recent decades is the development of new towns. The experience in India with such new towns has been moderately successful in a few cases, and not successful at all in numerous others. In general, such new towns rarely proved to have a significant impact on the regional patterns of urbanisation.

New Bombay, or *Navi Mumbai*, is one such an example of new town development in India. It is India's largest and most significant urban planning experience since Independence, which is increasingly being used as a *model* and adopted elsewhere in the country. Located opposite the Bombay peninsular on the mainland New Bombay has, since the early 1970s, been planned to become an independent twin-city of metropolitan size, with the aim to reduce the urban and economic pressure on Bombay.

This book is a slightly modified and concise version of an academic effort that was started at the School of Oriental and African Studies (SOAS) of the University of London in 1994, and resulted in a PhD thesis four years later. Although the book does not provide all the data that were included in the dissertation, hard copies of the original work can be found in both the SOAS and the Senate House libraries of the University of London.

The major aims of the book are (1) to evaluate the urban planning experience in New Bombay and to learn some lessons from it, so as to uncover the potential theoretical implications of the New Bombay model for urban development in the Third World in general; and (2) to provide an independent pioneering study, largely based on empirical research, on the newly developing twin city of Bombay.

The book consists of three parts. In Part I the focus is mainly on urban development processes and patterns, starting from the broadest level (the developing world) in chapter 2, and narrowing it down to India (chapter 3) and Bombay (chapter 4). Part II contains the major part of the data and the results of the fieldwork in New Bombay. The focus has been on (a) the development concept and planning objectives; (b) the extent of growth and development of New Bombay after 25 years of project implementation; (c) the physical and socio-economic impact of urbanisation and development on the original villages and the village population; (d) the extent to which the new town is catering to all social classes; and (e) the role of the State in the dynamic process of socio-economic change, with the focus primarily on the urban poor. In part III, New Bombay's planning concept and plan implementation have been further discussed. In general, the book's focus is more on the socio-economic aspects of the urban planning *model* that was adopted in New Bombay, rather than on the physical aspects of urban planning.

1 Introduction

' In an already largely urban world the growth of cities
will be the single largest influence on development
in the first half of the 21st century. '

(United Nations Population Fund, 1996)

India in general, and Bombay in particular, are some of the world's most dramatic examples of Third World urban growth. In 1941, 43.6 million Indians were living in urban areas, which was 13.7% of the then total population of 318.7 million. By 1991, the number of Indian 'urbanites' had increased to 217.6 million, which was 25.7% of the total population of 846.3 million in that year. Within the same fifty-year period, the population of Bombay, a city which 350 years ago did not even exist and is now the sixth largest city in the world, increased from 1.49 million in 1941 to 12.60 million in 1991.[1] Within that 350-year period the city developed from a minor trading post on the west coast of the Indian subcontinent into the most diversified and economically, politically, and culturally most important city in the West Indian region. Today, Bombay is India's largest city in terms of population, and the country's premier commercial, industrial and financial centre.

New Bombay, or *Navi Mumbai*, is the twin-city of Bombay, which was being planned and developed from the early 1970s to bring some relief to the urban problems of the old city. *New Bombay* is by far India's most significant urban planning experience, and the country's largest urban development project ever undertaken, with a very significant regional and national impact in terms of urban planning policy. In this book we will examine the processes of urban development in *New Bombay*, against the background of urban development in a wider Third World context.

1

Urban development in its broad context

World urbanisation

In terms of global trends in society, the twentieth century may be remembered as the century of the urban transition; a transition from a predominantly rural society to an urban society. Over the last ten to fifteen years, however, the processes of world-wide urbanisation and rapid urban growth have generally come to be considered as some of the major problems and challenges of the coming decades. This increasing consciousness has found expression in numerous workshops, seminars and conferences, the biggest and most recent one, the United Nations' *International Conference on Human Settlements* (Habitat II), held in Istanbul in June 1996.[2]

Whereas in 1800 not more than 25 million people (or about 3% of the world's total population) were living in what were then called *urban areas*, by 1950 that number had gradually increased to 300 million (about 12.5% of world population), and only one city in the world (New York) could then be called a *mega-city*, having a population of over 10 million (with Greater London close to joining this category). From then, however, within an extremely short time span of 45 years, the world's urban population has exploded, from 0.3 billion in 1950 to 2.6 billion by 1996 (about 45% of world population), and 14 cities are now being called mega-cities (UN, 1996:70). By 2015, 16 years from now, the United Nations believe that the world's urban population, which is growing at a current rate of 61 million a year, will be over 4.0 billion (which will then be 54% of world population), and 13 more cities will have joined the 'mega-city league'. By that time, even the 15th largest city in the world will have a population of over 15 million.

This dramatic pace of world urbanisation since the 1950s has largely been the result of urban growth in developing countries. In most western nations urbanisation started with the Industrial Revolution in the 19th century, came to a height in the early 20th century, and was by the 1950s already slowing down and gradually stabilising.[3] By 2020, when the world's urban population will have increased by 2.06 billion compared to 1970, nearly 93% of this increase will have taken place in urban areas in developing countries, the 13 new mega-cities mentioned above all being Third World cities (UN 1996:29). By that time, 75 to 80% of the world's total urban population will live in Third World cities.

Figure 1.1: World population growth and urbanisation 1800-1996, with 2015 projection (in millions)

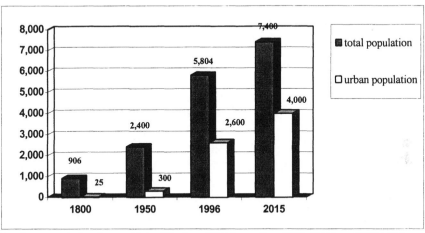

Based on population figures United Nations 1978 (pop. figs. 1800, 1950) and UN 1996

Figure 1.2: Number of megacities (10 million+) in western and developing countries 1950-1994, with 2015 projection

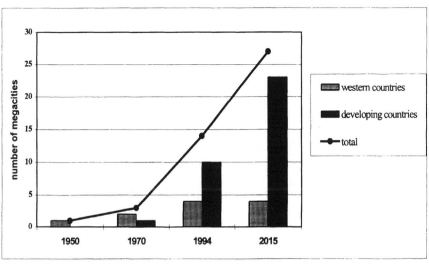

Based on figures United Nations 1996

Although the processes of urbanisation and rapid urban growth in the developing world have many similar features as the same processes in the western world several generations earlier, there are some major differences.

First, in the western world rapid urban population growth coincided with, and was a direct consequence of the industrialisation process and increasing urban prosperity. This caused a demographic transformation with, on the one hand, the urban population, due to a large urban migration process, rapidly increasing and the rural population decreasing. On the other hand, it resulted in an overall slow-down in population growth. In the developing world of today, industrialisation is only one of many factors responsible for the urban migration process, and in most Third World countries not even the most important one. Moreover, in most large cities of the developing world urban migration has, in terms of overall city growth, become less important than urban natural increase (Preston, 1988:11-31, World Bank 1991:200, UN 1996:35).[4] This is a crucial difference from what happened in the western world.

A second and more important difference lies in the sheer numbers of people that are involved in the process. Taking India as an example, today just about 27% of the country's population is living in cities,[5] and urban growth over the last one and a half decades was only just over 3% p.a. (UN 1996:71 and Tata 1995:48). Comparing both figures with the earlier urbanisation process in the western world, these figures are relatively low. The proportion of urban population to total population was higher in the industrialising world of the late 19th and early 20th centuries than it is today in many Third World countries.[6] Furthermore, over the last decades, overall urban growth figures in developing countries have only in a number of nations been higher than they were in the western world at the height of the Industrial Revolution. Indeed, if the number of international migrants that left the UK for the US in the late 19th and early 20th centuries had gone to British cities instead, urban growth figures in Great Britain would have exceeded 5% p.a., which is a figure that is comparable to the annual urban growth in the developing countries as a whole, which peaked at 5.1% during 1955-1960 (UN 1996:26). Nevertheless, the enormous population figures of some developing countries imply that the absolute number of people that is involved in the process far exceeds those of the industrialising western countries earlier.[7] Due to the enormous population figure of India, for example, the country's relatively low urban growth figure, non the less, implies that 257 million people in India are currently living in cities. This number more or less matches the *total* population of the United States. In fact, there is only one country in the world (China) that has a larger *total* population than India's urban population!

Looking at growth figures of cities separately, the majority of cities in the developing world have figures that are sometimes comparable, but sometimes also considerably higher than those of the industrialising cities in Western Europe and the US earlier. Taking Bombay, one of the world's largest and fastest growing cities over the last five decades as an example, barring the decade 1941-1951 the city's urban growth figure never exceeded 4.5% p.a.[8] But, again, the absolute numbers involved were in Bombay of a complete different magnitude than they have been in any western city. Within a fifty-five year period, Bombay's population increased from 1.49 million in 1941 (Census of Maharashtra 1991) to around 15.0 million today (estimate).[9]

Already by the end of the 1970s, the less developed countries had overtaken the developed countries in terms of urban growth. From that time, the developing world counted more 100,000+ cities, more 1 million+ cities, more 5 million+ cities and more 10 million+ cities than the developed western world. Most significant is that in 1950 twenty of the thirty largest cities in the world were western cities. By 2015, their number will be reduced to five, and no less than twenty-five of the thirty largest cities will be Third World cities.

The magnitude of the present phenomenon of worldwide urban growth, especially in the developing countries, has led many observers and specialists to describe the current situation, in terms of the impact that it will have for the history of civilisation, as nothing less than a *Third Revolution*, in succession of the agricultural and industrial revolutions (Brunn and Williams 1983). According to the UN, this dramatic urban growth will be 'the single largest influence on development in the first half of the 21st century' (UN 1996:1). In chapter 2 we shall discuss the urban development processes in the Third World in greater detail.

Consequences of the urbanisation process

Rapid urbanisation in Europe and the US, especially in the late 19th and early 20th centuries, caused many problems. In general, proper urban infrastructure and basic urban services were lacking, living conditions were appalling and millions suffered from a lack of hygiene and from all sorts of diseases. The overall urban situation in Manchester in the 1830s for instance was probably, at least for a considerable number of people, not very much different from the situation in many developing countries today. The difference with the current urban growth in most developing nations, however, is that the ratio of capital generation to population growth has been much higher in the industrialising western world than it is in the post-colonial developing world. For Third World cities this has consequences on three levels: in economic terms it has created widespread unemployment and underemployment, and mass poverty has become one of the main

characteristics of city life; in social terms almost all cities are having massive problems providing sufficient infrastructure, housing and services to their citizens, who are also physically and mentally suffering from environmental degradation, lack of open space, etc.; at the political level, local government and administration are too often characterised by widespread inefficiency and loss of real power, and are under constant pressure from all levels of urban society.[10]

Over the last two decades, housing and urban infrastructure have generally been considered as the two major problem areas of cities in the developing world. Until recently, the major international development agencies, as well as national and regional governments in developing countries, considered urban migration, quite correctly, as the main cause of rapid urban growth. Therefore, the policy guidelines, over the years, were largely focused on measures to keep these migrants away from the cities. Rural development, industrial decentralisation and deconcentration, and the creation of satellite towns or so called *new towns*, have in almost all developing countries, in one combination or another, been adopted. However, as mentioned before, urban growth in most large cities in the developing world is at present more the result of natural increase than of urban migration, and secondly, as economic factors are not the only reason for rural people to migrate to cities, rural development alone has rarely prevented people from migrating. Consequently, the impact of such policies on the overall problem of rapid Third World urbanisation has been minimal.

From the early 1990s, and at the instance of the United Nations and the World Bank, the urban development policy framework was therefore completely revised, and attention was shifted from rural development towards *urban growth management*. The great importance of Third World cities as the economic engines of their respective countries was duly acknowledged and became the central focus of economic planning. Trying to slow down or to prevent further urban economic growth was considered much worse for the country and its people in general than the problems caused by urbanisation (UNDP Strategy Paper 1991, World Bank Policy Paper 1991). Moreover, further urban growth was not only considered as a powerful tool for economic development, but also for social development, as many countries in Asia and some in Latin-America have illustrated. Indeed, indicators of health, social mobility and education for example are all much higher in urban areas.

Whereas increasing urban growth has the potential for improving the living conditions of millions, the ultimate challenge will be to link urban social and economic development to the problems of unprecedented growth and urban constraints and congestion. Although further urban growth may indeed pose serious threats to the urban environment in terms of collapse of basic urban services, intolerable environmental degradation and escalating social conflict, these large cities nevertheless carry in themselves the potential to create economic, cultural and social progress; they must be recognised as instruments which can open up new avenues of human development.

Choice of case study

With India presently having an urban population of around 260 million (coming from just over 43 million in 1941), and the Greater Bombay Urban Agglomeration presently having an estimated population of around 15 million (about ten times as much as in 1941), both India and Bombay are some of the world's most dramatic examples of Third World urban growth. Given these tremendous urban growth figures it is rather surprising that there has never been an overall urban policy framework from the Indian federal government for the urban areas in the country at large.[11] The urban policies followed and implemented by India's major cities, have been *ad hoc* policies, differing from state to state, and differing from city to city as well as from period to period.

The sharp increase in population in the city of Bombay, within such a relatively short period, is one of the clearest reflections of the city's economic success. Unfortunately, at no point in the post-Independence period have the local and State authorities been able to manage and direct this rapid urban growth. The rapidly increasing population has put more and more pressure on the urban infrastructural system and, vice versa, severe urban congestion and inadequate basic urban services have put increasing pressure on both the city's population and the local economy.

In an attempt to avoid an urban collapse, which at the time was considered as quite possible by most people, the Government of Maharashtra, after years of discussions and deliberations, decided in the early 1970s to create a large new *twin-city*, opposite the Bombay peninsular on the mainland. This New Bombay *or Navi Mumbai*, as it was called, was to become a self-containing 'counter-magnet' of metropolitan size, was to relieve Bombay of the worst urban pressures, and was to bring some corrections and balances to the city's growth pattern. The new town had to become a well-planned and pleasant area to live, had to create its own economic base, had to attract some of the existing population of Bombay and to divert new in-coming migrants away from Bombay, was to fully integrate the existing local population into the newly developing area, and was last but not least to become a city in which all people, rich and poor, would be able to find a place to live. A quasi autonomous state agency was to be established to implement this New Bombay plan, which was also to be self-financing. The agency had to acquire all the land that was necessary, had to undertake all development work, provide the necessary social and physical infrastructure and had to recoup the entire cost of development from the sale of land/housing and structures, which could then be re-used for further development (for more on this see chapter 5).

Thus, although the New Bombay plan was broadly based on the *New Town concept* which was originally developed in Britain in the beginning of this century, and was later adopted elsewhere in Europe, the US and the former Soviet-Union, in terms of geographical location, project objectives, general planning and development concept, and financing strategy, the New Bombay plan was in those days a fairly unique initiative.

Aims of the study, book structure, survey research and data collection

Aims and nature of the study

The main purpose of the study is to examine the processes of urban development in New Bombay, India's largest and most significant urban planning and development experience since Independence. Naturally, experiences with urban planning and development are not restricted to the local level, but may have, depending on the overall success, a very significant regional and even national impact. In the case of New Bombay the impact has indeed been very significant. The planning agency that has been in charge of developing New Bombay was later also designated the planning authority for numerous other and similar urban development projects in the state of Maharashtra and outside. As a result, the urban development 'model' that was adopted on a fairly experimental basis in New Bombay is now also being implemented elsewhere in Maharashtra and India; a process which may have an ever-lasting imprint on India's urban future. Furthermore, the New Bombay *model*, if successful, may also provide valuable elements of Third World urban planning to other countries facing similar problems of dramatic metropolitan growth and urban congestion. It is, thus, the lessons from urban development in New Bombay and their possible use elsewhere that is at the core of this study.

Going by the official statements and reports of the development authority, and looked at from outside, the New Bombay project seems to be a major success. The objective of this book is to examine the extent to which the implementation of the New Bombay project, given its specific goals and based on a specific development and financing strategy, can indeed be called successful. To this end, we will first examine the extent of success of the New Bombay model in terms of the physical and economic growth and development of the twin-city, and its possible effect on the decongestion of Bombay. Secondly, we will also examine the extent of success in terms of social development in the twin-city, whereby the major focus will be on the impact of the New Bombay model for the urban poor.

Few subjects in science require such a wide interdisciplinary approach as the study of urbanisation and urban growth, and the numerous topics related to it. Apart from geographers, economists, political scientists, anthropologists, sociologists and other social scientists, topics related to urbanisation and urban growth are also being studied by urban planners, architects, engineers, and so on. A study focusing on the urban planning of a new town and its economic and social implications must therefore necessarily be broad in scope. Consequently, although several aspects of urban planning and urban development in New Bombay have been studied in-depth, the overall approach of the topic has been kept deliberately broad.

If it is at all proper to make a distinction in this kind of work between a *theoretical study* and a *practical study*, this work belongs to the second category. Whereas analysis of the New Bombay model will uncover the potential theoretical implications for urban planning, both physically and socio-economically, the major aim of this study is to be a possible functional and practical tool, to urban development authorities and other interested parties in the developing world.

Besides a theoretical contribution to urban planning theory in general, the study will also significantly contribute to the knowledge of the case, as it will provide large sets of specific data on physical and economic growth in New Bombay, as well as on social processes in the new town. So far, very little independent research has been done on New Bombay, and another major objective of this book is therefore to function as a pioneering study on New Bombay. Logically, therefore, the major part of the data included in this book will be based on emperical research (see further).

The specific nature and scope of the study does not allow for one central hypothesis as the starting point of this research. This study starts from a number of questions, which are largely open. Subsequently we shall, on the basis of the collected data, evaluate the urban development processes in New Bombay, and formulate answers to the questions that have been put forward. They should provide some insight in the urban planning and development practice in New Bombay, and should provide some valuable new elements of, and new insights in urban planning theory in general.

Finally, it has to be clarrified and stressed from the very beginning that it is not the intention to place the discussions in this book within the classical theoretical framework of ideologically opposing viewpoints, as is sometimes thought necessary. Too often, academic argumentation systematically starts from an ideological point of view, whereas a truly independent scientific approach ought to start, as Henri Poincaré already argued, from the plain facts instead of an ideology. Elements of theory, which may be called *Marxist* in some instances and *Liberal* in others, may both help to explain that reality, as Galtung already argued.[12] Throughout the process of data collection, and throughout this book, these research principles have functioned as the major research guidelines.

Structure of the book and survey research

Part I of this work will discuss very broadly urban development and planning processes in the Third World in general, theoretically as well as practically (chapter 2), and will thereafter more specifically examine the urban development processes in India (chapter 3) and in Bombay (chapter 4), the latter predominantly in historical, demographic and socio-economic terms. Part I will therefore provide the context for the study of urban development processes in New Bombay.

In part II the major results of the survey research will be presented; work that has been carried out in New Bombay over a period of about a year. The origins of the twin-city and the initial policies that were adopted and strategies that were planned will be discussed in chapter 5. The three remaining chapters of part II will all focus on the extent to which New Bombay has developed economically and physically, and in which direction the new town has developed in social terms. The fieldwork was planned as to provide data to analyse: (i) the extent of economic and physical growth and development of New Bombay after 25 years of project implementation (chapter 6); (ii) the impact of urban development in New Bombay on the original population in the villages (chapter 7); (iii) and the socio-economic development in the new town (chapter 8). A discussion and analysis of these fieldwork data, and of the processes behind them, will form a significant part of these three chapters.

In part III both the initial planning concepts and the overall plan implementation will be further analysed and critically discussed. An attempt will be made to assess the ultimate regional impact of the twin-city on the old city of Bombay (chapter 9), and in the final chapter of the book (chapter 10) the major conclusions will be presented, as well as the scope for further research on the topic.

Data collection and methodology

The Indian government has been organising decennial Census surveys since 1871, which are commonly considered as to be quite reliable. As a result, a large body of literature, academic as well as non-academic, is based on such Census data and is readily available throughout the country. In Bombay also there is no shortage of data and the number of publications on the city is fairly large. New Bombay, on the other hand, is a newly developing area and has so far rarely been the focus of independent academic interest. The number of people that have been looking at urban development in New Bombay from an academic point of view is very limited. Therefore, this study is predominantly based on primary sources and empirical research, and aims to be a pioneering study on urban development in *Navi Mumbai*.

As this research does not look at one small research topic but at a whole set of research items, the procedure of data collection in the field has been different from survey to survey, dependent on the kind of data needed.

Survey techniques that have been used include policy analysis, RRA/PRA, questionnaire surveys, formal interviews, and informal discussions.

First, data concerning the origins of New Bombay (chapters 4 and 5) and those concerning the formal planning and development policies and practices that were adopted (chapter 5) have been collected primarily from the New Town Development Authority (NTDA) and other state government agencies (BMRDA, Urban Development Department). Other data that have been acquired through *deskwork* are the statistical data on the extent of physical and economic growth and development in the area (chapter 6). Most of these statistics were available with the NTDA in some form or another, and were brought together bit by bit from different offices and desks.

Secondly, the impact of urban development in New Bombay on the original population in the villages (chapter 7) has been examined by using Rapid Rural Appraisal (RRA) (and to a lesser extent Participatory Rural Appraisal, PRA) as the main survey technique. RRA has also been used to double-check the questionnaire data collected in the *sites and services* of New Bombay (the results of which have partly been included in chapter 8). Since RRA and PRA are 'modern' and not widely used survey techniques, a concise discussion on the origins, the nature and the value of these techniques will be provided in chapter 7.

Thirdly, to analyse the main elements of socio-economic development in the twin-city, it was planned to conduct a large questionnaire survey in the different newly developed nodes and in the sites and services of New Bombay. However, shortly after my arrival in New Bombay the NTDA had a similar and obviously much larger questionnaire survey in the pipeline, to be conducted in all the different areas of New Bombay (viz. the nodes, the villages and the site and service areas). The questionnaire survey had a random sample of 25% of the total population in 1995, i.e. a sample of nearly 20,000 households, or the equivalent of 80,000 people. For more details about the set-up, the value and the results of the questionnaire I refer to chapter 8. The data in chapter 9, which give an idea of the extent to which New Bombay has become a 'self-contained' city, are also based on the same questionnaire survey.

Finally, formal interviews with key-people, from government agencies as well as with non-officials, and regular informal discussions with people specialised in specific areas, have contributed significantly to gaining a basic insight in the developing processes in New Bombay.

Some preparatory notes

The terms used in the literature to describe the economically less developed countries of Asia, Africa and Latin America, are numerous and quite diverse, ranging from *Third World*, over *developing countries* or *developing world* to simply *the South*. The UN usually groups these nations under the heading *LDCs* (or Less Developed Countries), as opposed to the *MDCs* (More Developed Countries). With the recent collapse of the Soviet-Union and other socialist countries in Eastern Europe, the term 'Third World' has lost its original meaning of the 1950s and the decades thereafter. Moreover, with the ideological-economical shift towards capitalism and the economic integration in the world economy of a large number of developing countries, a major part of this 'Third World' can no longer be considered non-aligned, and some regions within this Third World are now as highly developed as some parts of the First and formerly Second World. In the text hereunder we shall employ the term 'Third World' in its non-ideological interpretation, as to plainly refer to the group of economically less developed nations of the South.

From late 1995 onwards, the newly elected BJP-Shiv Sena Government of Maharashtra gradually started to change the English street names as well as the names of numerous railway stations and buildings in the city for a new (and sometimes unpronounceable) Marathi version. The very city names of Bombay and New Bombay were also changed into respectively *Mumbai* and *Navi Mumbai*. Most organisations, corporations, etc. that carried the name *Bombay* have changed this to *Mumbai*. Hence, the BMRDA is now called the MMRDA. Other bodies, mainly at the municipal level, have kept their old names in order to avoid too much of confusion, but have given an addition to it. Throughout this book however, we will, for reasons of recognition, continue to use the former English names.

For so far as the data collection in the field is concerned the cut-off point in this book is 1 May 1996. For the more general data the cut-off point is 1 May 1997.

To have an idea of price levels mentioned in the text, the exchange rates between British Pound and Indian Rupee over time were as indicated below. In India one speaks of *lakhs* and *crores* of Rupees: one lakh is Rs. 100,000 and one crore is Rs. 10 million. In the period of fieldwork, between November 1994 and March 1996, Rs 100 equaled £2.08 to 1.82, in June 1996 Rs 100 was worth £1.86, in January 1997 it was worth £1.67, in June 1997 £1.70, and in January 1988 £1.50.

Finally, the abbreviations that have been used in the text have been listed at the beginning of the book, and those appendices that were considered as most relevant can be consulted at the back.

Notes

[1] The 1991 figure obviously refers to a much larger geographical area.

[2] Some other important urban conferences and meetings of the past decade are the International Conference on Population and the Urban Future (Barcelona, 1986), the Asia Conference on Population and Development of Medium-sized Cities (Kobe, 1987), the international workshop on Cities of the 1990s, organised jointly by the Development Planning Unit (University College London) and the ODA (London, 1991), and the International Colloquium of Mayors on Social Development (New York, 1994).

[3] At the beginning of the 20th century only one nation, Great Britain, could be regarded as an 'urbanised nation', with a majority of its population living in urban areas. The US would become urbanised by 1920. By 1955 there were at least 18 urbanised nations.

[4] (a) For definitions of urban growth and urbanisation see chapter 2; (b) Part of the African continent is an exception to this general observation.

[5] In part this is, of course, a result of the constantly increasing population in rural areas.

[6] At present, the African continent as a whole is for only 34% urbanised and the Asian continent for 35%. Europe, North America and Latin America, on the other hand, are each having an urban population of 74 to 76% (UN 1996:70-72).

[7] For example, the total population of the entire European continent in 1901 was 401 million, or less than half of India's total population today (953 million) (Bradnock 1989:table 2 and UN 1996).

[8] Not many of the large Third World cities experienced higher growth rates than Bombay. Karachi, Lagos and Dhaka have been growing by 4.7, 6.7 and 7.6% p.a. respectively between 1970 and 1990.

[9] The estimate is based on the official figure for 1991 and on Bombay's average growth rate of 3.70% p.a. during the period 1961-1991.

[10] Rapid urban growth, as it is in part the result of movements from countryside to city, obviously has also many socio-economic consequences in rural areas (changing social organisation, financial transfers, transfers of urban values).

[11] The only attempt to such a policy framework was caried out in 1975 (see ch. 3).

[12] Galtung's 'Principle of Variety of Theories' in his Theory and Methods of Social Research (1967).

PART I

URBAN DEVELOPMENT AND URBAN PROBLEMS
IN THE THIRD WORLD

2 Urbanisation, Urban Growth and Planning in the Third World

' The tremendous expansion of cities, especially in developing countries, is transforming social dynamics throughout the world. But while it brings daunting challenges, it also brings unprecedented opportunities.'

(Boutros Boutros-Ghali, 29 May 1996)

Introduction

In this chapter we shall first have a look at the urbanisation and urban growth processes in the Third World in general, their regional differences, common characteristics with the western experience earlier, the different factors contributing to urban growth, and the impact of colonialism. Subsequently, we will discuss the main theories of urbanisation and we will have a look at the emergence of the Third World metropolis. Separate sections in this chapter are reserved to discuss the major problems facing Third World metropoles, and the strategies that have been followed, at the national level and at the regional and local levels, in trying to overcome these problems.

     The terms that are being used in the extensive literature on urbanisation and urban planning are not only very diverse, they are also frequently used in different contexts or are interpreted quite differently.[1] Regularly, urbanisation is simply equated with urban growth, or the modern metropolis is considered as a modernised form of the industrial city. The arbitrary use of these definitions and concepts creates confusion and must be avoided. We shall therefore try to give acceptable definitions for the major terms that will be used in this chapter, as and when they are being discussed.

Urban growth and urbanisation in the Third World

Definitions

Urbanisation and *urban growth* are sometimes considered synonymous. They have, however, an entirely different meaning. 'Urbanisation in its most formal sense', writes Roberts, 'merely constitutes the increase of the urban population as compared with the rural one' (Roberts 1978:9). In this sense, urbanisation points to the '*relative* concentration' of people living in urban areas (Potter 1992:5). However, this demographic definition of urbanisation leaves some questions unanswered. For example, how to define this 'urban area'? In some countries an area is being called 'urban' only when it has a population of minimum 100,000, whereas in some other countries a population of 2000 is sufficient to be defined as 'urban' (Angotti 1993, Todaro 1984, Brunn and Williams 1983). Furthermore, shifts of administrative boundaries may transform a rural area into an urban area from one day to the other which, among other things, has far-reaching consequences in comparing data on urbanisation and urban growth over time.

This demographic definition of urbanisation is quite different from the 'larger' definition of urbanisation that is frequently used by Marxist writers in a socio-economic context, where it is related to the process of national and international economic transformation towards a capitalist form of production and consumption (Roberts 1978, Slater 1986, Lojkine 1976, Castells 1977). We shall come back to this later in this chapter.

Urban growth, on the other hand, points to 'the *absolute* increase in the physical size and total population of urban areas' (Potter 1992:5). Urban growth, therefore, is the sum and the result of three processes: net urban migration (in-migration minus out-migration) plus urban natural increase (births minus deaths) plus geographical extension of the urban area. In this sense urban migration may be associated with 'urbanisation'. Consequently, as urbanisation is one of the three major components of urban growth, urbanisation and urban growth are linked, but they are in no way synonymous. As Slater (1986:8) points out, the difference between the two terms may be reflected in policy guidelines. Urban growth which is largely the result of urbanisation may lead to a policy of economic decentralisation and rural development, whereas if it is largely the result of urban natural increase it may rather lead to a policy of population control.

Some figures

The following statistics illustrate the magnitude and speed of the urban growth process in the Third World, as introduced in the previous chapter.

Table 2.1 and figure 2.1 demonstrate that in the 45-year period between 1950 and 1995, the urban population of Africa increased from 32 million to 254 million, the urban population of Latin America from 67 million to 336 million and the urban population of Asia from 218 million to 1.12 billion (Todaro 1984:9 and UN 1996:70-72).

Table 2.1: Growth of urban population per world region (millions), 1925-2025

	1925	1950	1975	1995	2025
World total	*405*	*701*	*1548*	*2612*	*5341*
North America	68	106	181	225	313
Europe [a]	162	215	318	316	538
Asia [b]	103	218	574	1125	2718
Latin America	25	67	198	336	601
Africa	12	32	103	254	804

Source: 1995 figures and 2025 projection: UN 1996; all others: Pacione (1981) and Todaro (1984). Note: (a) Eastern Europe included in this figure; (b) Japan, the countries of the Middle East, and the countries of the former Soviet Union not included in this figure.

Figure 2.1: Growth of urban population (in millions), 1950-2025

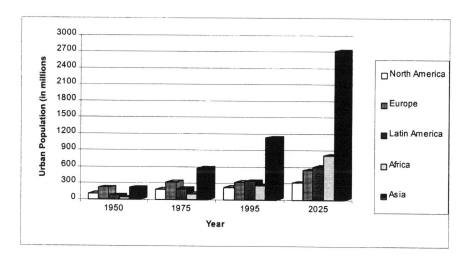

Thus, only forty-five years ago, when over half of the world's urbanites were living in cities in North America and Europe, the total population living in the urban areas of the Third World was about 317 million. Today, that number has increased more than five-fold to over 1.7 billion, and by 2025 it will have increased by another two and a half times compared to today and will be around 4.1 billion. By that time, only one in six of the world's urbanites will be living in cities in North America and Europe. The other five in six will be living in Third World cities (figure 2.2).

Figure 2.2: Proportion of urban population per region (1950 and 2025)

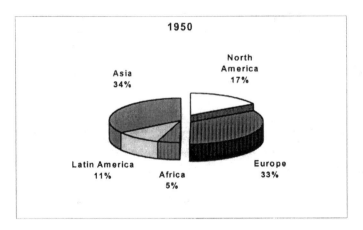

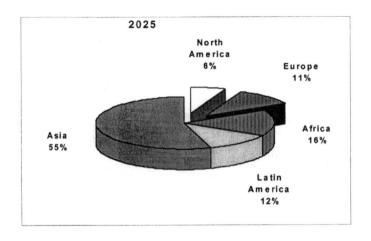

Figure 2.3: Urban population increase Third World, 1950-2025

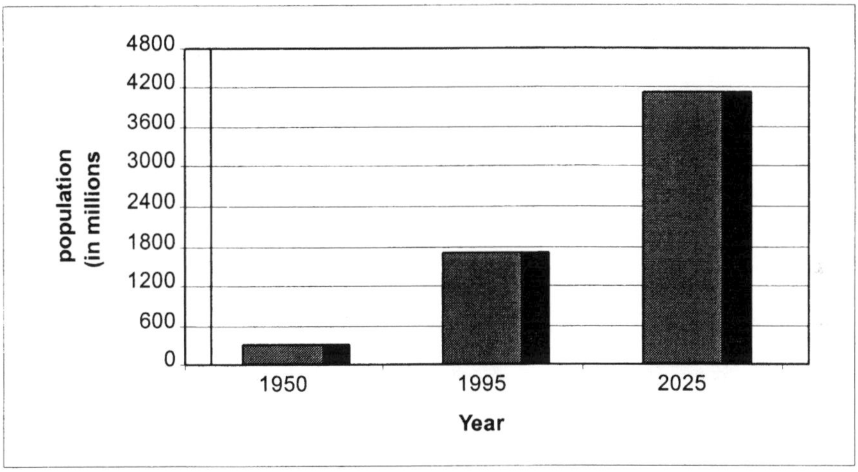

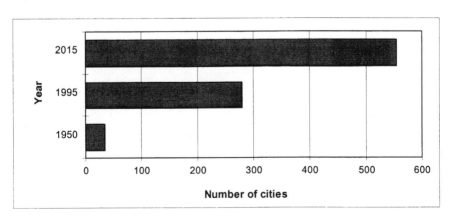

Based on figures table 2.1

In 1950, the Third World counted 34 so-called 'metropolitan cities' (1 million or more population). Today that number has increased to over 280 and is expected to almost double by 2015 (UN 1996:1).

Figure 2.4: Number of 1 million+ cities in the Third World, 1950-2015

In 1950, only 10 of the 30 largest cities in the world were located in developing countries, 3 of which in China and 2 each in India and Brazil. By 1990, the number of Third World cities in this list had doubled to 20. And by 2015, no less than 25 of the 30 largest cities in the world will be located in developing countries, 8 of which will be in South Asia alone![2] (UN 1996:32-33).

Finally, as has been mentioned in the previous chapter, it is most significant that all of the 13 cities that will join the world list of mega-cities between now and 2015 are all cities situated in the Third World (UN 1996:29).

These data clearly indicate that the rapid urban growth world-wide since the 1950s has almost exclusively been the result of the explosive growth of cities in the Third World alone, and that urban growth in the Third World is not expected to significantly slow down in the coming decades.

Natural increase and urban migration

Many large Third World cities, especially in Asia and sub-Saharan Africa, show a markedly unbalanced sex ratio, with a majority of the population being young males in the age category between 15 and 30 years old. This is an indication of a process of rural to urban migration in which young males move to the cities, the majority of them in search of a job and an income. Although some may only leave the countryside temporarily and move back to their rural homes from time to time, the large majority of them starts a new life in the city and is, in most instances, later joined by the family.

Urban migration is always motivated and therefore a result of *push* and *pull* factors, which may be situated in the economical, political, or socio-cultural sphere.[3] A poor agricultural labourer or petty land holder for example may for so many reasons be pushed away from his rural environment and pulled towards the city in search of a job in industry, or he may simply be attracted by the higher wages paid in the urban economy. Numerous studies give proof of the fact that such economic push and pull factors have always dominated the urban migration process (see for example Beier 1984, Gugler 1988, Yadava 1989). Nevertheless, major political events or decisions may be another reason for people to move to urban areas. When India was partitioned at Independence, several millions of Hindus moved from then West and East Pakistan towards the largest Indian cities, Bombay and Calcutta.[4] In Bombay for instance they ended up in large refugee camps in Ulhasnagar, at some distance of the metropolis, which after some years developed into one of Bombay's larger satellite towns. Higher education, better health facilities, or even the excitement of the big city, are examples of socio-culturally determined urban migration motives, as is the perception of the city as a place of refuge from one's culture or personal background. As Gugler indicates, in Latin America and Indonesia women frequently outnumber men in the urban population, as never-married, separated, divorced or widowed women are faced with limited rural options and are attracted by the cities that offer them better opportunities in terms of both work and (re)marriage (Gugler 1996:6). Finally, besides the economically, politically and culturally motivated urban migration, natural calamities in the countryside such as floodings, droughts or earthquakes may also push people *en masse* away from their rural homes into one of the larger metropoles. In general, as Beier simply concludes, 'people relocate because it is rational for them to do so' (Beier 1984:67). The possible motivations behind such a move are numerous and in most cases probably a mix of different considerations and the result of a careful weighing of the *pro's* and the *con's*.

In the 1950s and 1960s, urban migration was considered the most important factor in the growth process of cities in the Third World in general (Potter 1992). Since some time, however, urban natural increase has become markedly more important in most large cities. Yet there is some discussion on this hypothesis. Some indeed argue that urban migration has always been, and still is the most important factor in urban growth in a considerable number of countries in Asia, Latin America and Africa (Findley 1993:1-31, Renaud 1981:30). The majority of observers, however, fully agrees with the above hypothesis (Preston 1988:14-15, Slater 1986:8, World Bank 1991:200, UN 1996:27-35).[5] The discussion is mainly caused by the fact that children born in cities from migrant parents are by some considered as part of the *natural increase*, whereas others still classify them as *migrants*. Therefore, as Todaro has pointed out, 'while statistically the major portion of urban growth appears to stem from high rates of urban natural increase, in reality the most significant long-run contributing factor is rapid rural-urban migration' (Todaro 1984:20). To avoid confusion, in this book every urban resident who has been born in the city has been classified as such under the heading *natural increase*.

Moreover, as Beier observed, the importance of urban migration varies considerably from region to region and from period to period anyway, and depends to a large extent on the rate of urbanisation that a country has already reached (Beier 1984:67). Furthermore, if urban natural increase has become the more important component, this does not mean that the absolute influx of people has slowed down. It simply indicates that the urban population is already so large that the proportion of people moving into the cities is becoming relatively smaller compared to natural increase. Generalising conclusions should, thus, be avoided.

Often, urban migration is characterised by so-called *chain* or *step-migration*, whereby rural people migrate to the largest cities in stages, going from rural to small and local city, and from there to the larger urban areas (Potter 1992:13). According to Beier, however, it is becoming evident that migrants, due to the decline in transportation costs and to the presence of friends and relatives as sources of information and support, are increasingly bypassing any intermediate stops (Beier 1984:67). Here as well, however, significant differences occur between and within countries and regions.

Regional differences and patterns of urbanisation

Significant differences in urban growth rates exist between and within the regions of Asia, Africa and Latin America. In 1975, the Third World in general was for 28.0% urbanised. At that point, already 61% of Latin America was urbanised, but only 26% of both Asia and Africa. Percentage-wise, Africa experienced the highest increase of the three continents, but a relatively small increase in absolute size of urban population. Its annual urban growth rate reached a peak of 5.1%, which is the highest rate experienced by any region in the fifty years between 1950-2000. During 1950-1960, annual urban growth rates of Latin America and Asia were both 4.57%, and their rates have constantly declined since then (Todaro 1984:9-13). Today, 74% of Latin America, 35% of Asia and 34% of Africa is urbanised (UN 1996:70-72). For those countries with a large percentage of population already living in cities, the impact of migration on urban growth rates should decline rapidly, whereas in the mainly rural countries migration patterns will still dominate urban growth for some time to come.

These figures clearly indicate the different patterns of urbanisation throughout the Third World. Ashish Bose (1974) observed that the issues involved in the urbanisation pattern are very complicated. It is, according to Bose, not just a matter of investment in urban infrastructure or regulating the flow of migrants into the cities. Instead, the entire subject has to be understood in the wider context of economic growth, social and political change. Cohen classifies Third World countries in four different groups:

(a) countries in which the process of urbanisation is well under way, in which more than half of the population is urban and has relatively high incomes and in which there is little pressure on the arable land and natural resources; (b) countries with a recent urbanisation experience, in which over half of the population is still rural and in which pressures on land and incomes are relatively low; (c) countries that are predominantly rural, that are urbanising rapidly but will remain predominantly rural in the coming decades; and (d) countries with severe pressures on the land in largely rural, subsistence-level societies (Cohen 1984:29-30).

Beier has used Cohen's categories to classify most Latin American countries in the first category, the majority of countries of East Asia and North Africa in the second category, the majority of countries of sub-Saharan Africa in the third category, and most of the large countries of Asia, such as India, Pakistan, Bangladesh, Indonesia and China in the last category (Beier 1984:62-65).

These latter countries (China not included) would according to Beier remain dominated, at least until well into the next century, by large and growing rural populations living in absolute poverty, whereas at the same time their urban populations are already large and could swell massively. Lester Brown, in the mid-1980s, doubted whether the Third World urbanisation trends and prospects for the coming decades would hold. He put three arguments forward that could prevent further urbanisation in the Third World: insufficient food surpluses to feed the ever-growing urban (and rural) populations; insufficient cheap energy to be provided to the urban population; and insufficient productive employment in the urban areas. Brown therefore concluded that if urban migration cannot be slowed down 'some harsh correctives may begin to operate in the not too distant future (...) with food shortages in the cities, energy shortages that will hamstring the economy, and rising levels of unemployment' (Brown 1984:102).[6]

Urban growth processes in the Third World and the Western world compared

Historically, the nature of the current urban growth process in the Third World is quite different, quantitatively as well as qualitatively, from the one that occurred earlier in the so-called developed Western world. We already went into this very briefly in the introductory chapter. As Berry argued over twenty years ago, we must disavow the view that urbanisation is a universal process, a consequence of modernisation that involves the same sequence of events in different countries and that produces progressive convergence of forms (Berry 1973). Not only are we dealing, says Pacione, with several fundamentally different processes that have arisen out of differences in culture and time, but these processes are producing different results in different world regions (Pacione 1981). Some of the results of these processes (the preindustrial and industrial cities of 19th century Europe, the colonial cities, or the modern 20th century metropolis) are indeed entirely different in character and outlook.

As was already argued in the previous chapter, the main differences between the current urban growth process in the Third World and the one earlier in the western world were basically related to the demand and supply of urban labour and overall population growth. Urbanisation in the West was accompanied by, and was a direct consequence of a fairly gradual process of industrialisation and economic development, spread out over more than a century, which created a large demand for urban labour and brought relative prosperity to millions of people living in the cities. The dramatic demographic transformation from rural to urban areas, with in the initial phase little or no urban natural increase and an overall decline in population growth, was the direct consequence of these developments. Urban growth was, thus, almost exclusively based on migration and reached 2.1% p.a. on average in most European cities during their periods of fastest urbanisation.

In the Third World countries of today, urbanisation is only in part the result of industrial and overall economic growth. As Brunn and Williams point out, urbanisation in many Third World countries has mainly been the result of rising and unfillable expectations of rural people migrating to the cities to escape poverty and lack of opportunities. Urban migration in those countries was higher and took place within a much shorter period, with an average population increase in Third World cities in general of 4.5% p.a. in the 1940s and 1950s, but with many smaller urban areas growing much quicker than this.[7] Moreover, the process here was not accompanied by an overall decline in population growth, but by a continuation of such growth. Until today, overall population growth remains very high in both the urban and rural areas of the Third World, because of traditionally high birth rates and decreasing death rates.

Impact of colonialism

This brings us to an aspect that has been of great importance in the process of urban growth and development in the Third World. With the expansion of a mercantile society in Europe, the European pattern of urban development had extended to the European colonial areas. Under influence of the different objectives and interests of the colonisers (trade at first, later followed by industry), several older cities were transformed and new coastal trading posts spontaneously developed into thriving local economies. These developments had major implications for the internal pattern of urban development and lay-out of cities, as well as on the overall process of urban development and on the rural-urban balance.

In terms of urban lay-out, such new phenomena as the military cantonments, administrative towns, hill stations and colonial port cities were introduced, the latter which combined several functions and were most conveniently located along the coastal lines in order to serve as export bases for raw, mostly agricultural materials and resources and, some time later, as import bases for manufactured goods. Although in the initial phase of colonialism none of these new phenomena had much to do with industry, this was actually the time when urbanisation in many of the present-day Third World countries started. According to Potter, the existence of 'urban primacy', the growth of one or several very large cities, is an indication of this process and is one of the typical phenomena of former colonial countries, especially of those which are small, have low incomes and have agricultural economies geared to export.[8] With the industrialisation in Europe and the emergence of the European 'industrial city', which was obviously 'qualitatively distinct from pre-industrial development' (Angotti 1993:16), many colonised towns and cities of the Third World also shifted more in the direction of industrialisation, and increasingly started to characterise similar features as the industrial cities of Europe.[9] By the time that these colonial towns and cities experienced their major urban boom in the post-colonial era, they were however no longer merely industrial cities but had become real metropoles, with industry being only one of several sectors of the urban economy.

Even more important in terms of rapid post-colonial urban growth was the impact that colonialism had on the traditional social and economic structures. Although there are significant differences from country to country, in part as a result of differences in colonial policy and strategy of the European powers, in most colonised countries these structures were turned upside down and were drastically reformed, again with the sole purpose of political control to serve the European economic and financial interests (see for example Gilbert and Gugler 1984:11-26). The subtle socio-economic balances that existed in the countryside were in many places destroyed, for example by shifting land ownership into the hands of a few and creating land monopoly systems. In many places rural poverty became a new fact of life which, after independence, would become one of the major contributors (the *push* factor) to urban growth in Third World countries.

Theories of urbanisation and emergence of the Third World metropolis

Modernisation theory versus dependency theory

The literature on current urbanisation processes and patterns is shaped by the historically divided interpretations of development theory in general between modernisation and dependency theorists.[10] *Modernisation theory* traditionally follows the ethnocentric, paternalistic view of western universalism, that change from traditional to modern in Third World countries was to occur through the diffusion of capital, technology, values, institutional arrangements and political beliefs from the West to the traditional societies. The theory points to factors within the national framework, such as authoritarian and military political regimes or corrupt bureaucracies, as the main cause of persistent underdevelopment in the post-colonial Third World. *Dependency theory*, which developed in the late 1960s as a reaction to modernisation theory and was first developed by Cardoso and Faletto for Latin America and Andre Gunder Frank for the Anglo-Saxon world, starts from the larger international context and explains poverty in the Third World as, historically, a result of colonial exploitation. Although these countries have become politically independent, dependency theorists argue, economically they have been kept in a tightly dependent position within a global capitalist economic system.

Clearly, in explaining the underdevelopment of the Third World both arguments have some value. In most developing countries underdevelopment has no doubt historical external roots (colonialism), but that does not mean that more recent internal factors (military rule, bureaucracy, corruption) are of no importance. As Potter rightly argues, one thing that can be said with certainty is that 'the evolution of the developed and less developed areas of the world has been very closely interrelated over the past five hundred years' (Potter 1992:14).

The same theoretical divide between modernisation and dependency theory can be found in the discussion of the role that cities play in the overall development of the Third World. Modernisation theory argues that towns and cities create and spread development, and that development therefore 'should come through the positive diffusing influence of the city acting on the countryside as a catalyst of transformation' (Slater 1986:9). More or less the opposite view persists in dependency theory which claims that urban regions, since they are closely linked to the international capitalist system, grow by exploiting their peripheries. When the debate was raised in the early 1950s, the predominant view (defended by Hoselitz, Hirschman and others) was that cities indeed played such a generative role in development. It was therefore also believed that the creation of new urban entities (new towns, growth poles, growth centres) would contribute to and create economic development in the surrounding regions. A lot of criticism against this rather optimistic view on development came for example from Castells, Quijano and Santos for the Latin American area and from Roberts, Myrdal and Safa for the US and Europe.[11] The Myrdal school of thought argued that economic development in urban areas only created further growth in the cities themselves and their adjacent areas, and that in general the existing inequalities between urban core and rural periphery, or so called 'urban-rural duality', only increased. Safa observed that 'the different patterns of urbanisation in the Third World stem from their late entry into the global capitalist economy and dependence on advanced industrial societies for capital, technology, export markets, etc.' (Slater 1986:12). In the same fashion, in their 'Theatres of Accumulation' Armstrong and McGee argued that the urban areas of the Third World had become centres of production, without there being any spread effects to rural areas. In terms of consumption, on the other hand, these spread effects were clearly present. This pattern of concentration of production and spread of consumption would, according to Armstrong and McGee, hold back any rural development.

Today, it can be safely argued that such inequalities between urban growth and rural backwardness are not just a temporary event on the way to an even pattern of modernisation and development, and that in many Third World countries, to put it in Rondinelli's words, 'the primate cities have become islands of modernisation' (1984:218). On the other hand, it can not be argued that there are no spread effects into rural areas at all. Surely, urban centres do have certain positive effects on rural development as well.[12]

Growth of the Third World metropolis

Another field in which the dependency theory has had a significant impact, is in explaining the emergence of the Third World metropolis.[13] Today, not less than 40% of all metropoles in the world are located in Asia alone.

The modern metropolis became an international phenomenon in the 20th century, in developed and developing countries, in capitalist, socialist and mixed economies. As Angotti stresses, it is important to distinguish between the metropolitan and non-metropolitan settlements, particularly between the metropolis and the industrial city of the past. Most of the western industrial cities of the 19th century did not exceed a population of several hundred thousand and would today rather be considered relatively small towns. Some of these towns disappeared when the local industry collapsed or moved out. The modern metropolis, on the other hand, of which Jean Gottmann said that it was 'the largest and most complex artefact that humankind has ever produced' (Angotti 1993:1), is a relatively stable and durable settlement form. It does not show signs of loosing in importance because it does not depend on a single industry or economic activity. It reflects, as Angotti argues, 'a new and complex integration of economic activity, including industry, commerce and services throughout society' (Angotti 1993:2).

Angotti finds it surprising how few urban theories make a clear distinction between industrial city and metropolis. 'It is common for scholars', he says, 'to describe the transformation from industrial city to metropolis as only a gradual evolution, and blur the revolutionary aspects of the transformation' (Angotti 1993:7). His argument is mainly based on Hans Blumenfeld's writings of the early 1970s, who identified the universal characteristics of the metropolis, but never looked very closely at the particularities of the Third World. In 'The Modern Metropolis', Blumenfeld stated that 'from its long, slow evolution the city has emerged into a revolutionary state. It has undergone a qualitative change, so that it is no longer merely a larger version of the traditional city but a new and different form of human settlement' (Blumenfeld 1971:61; see also Kopardekar 1986:32-35). Not only is the population of this modern metropolis up to ten times larger than the industrial city, and is the land area up to 100 times larger, the most significant economic and social characteristic that distinguishes the metropolis from other settlement forms is functional diversity and specialisation, which is fuelled by technological innovation and stimulates a more complex division of labour.

Angotti also stresses the importance of recognising the distinct characteristics of metropolitan growth and planning in the US, Europe, the former Soviet Union and the developing nations of the Third World. 'The images evoked by the metropolis in each case are different, the problems are different, and the approaches to planning are different. While there are universal laws of metropolitan development that shape all of them, there are significant particularities that distinguish them' (Angotti 1993:29). Angotti identifies the US metropolis as the classical expression of 20th century capitalist urban development. It is an unequal metropolis based on the segregation of land uses and the fragmentation of social groups, it has a densely developed Central Business District (CBD) and sprawled suburbs, its population is highly mobile and depends largely on private transportation, and the presence of planning is very weak. The Soviet metropolis, on the other hand, was for Angotti the classical expression of 20th century socialist urban development, characterised by a relatively integrated social and political structure with limited social mobility, the presence of an administrative/residential centre and relatively high density suburbs and mass transportation. Planning followed a centralised administrative structure and tended to reproduce centrally-determined plans. The European metropolis shows characteristics of both, says

Angotti, as it is in general an expression of a mixed economy, but with more and more elements of the US metropolis. European metropoles are smaller, more compact and less sprawled, and tend to be more integrated in both social and land-use terms than the metropolis in the United States.

The Third World metropolis, on the other hand, mirrors according to Angotti the development of both the cities of the former colonial powers as well as that of the US metropolis of the 20th century, being the predominant metropolis and urban planning example of this century. He calls the modern Third World metropoles 'dependent metropoles', referring to the debate between modernisation and dependency theories, which are defined as 'metropolitan areas whose economies depend heavily on the developed capitalist countries, and which usually rely on export industries' (Angotti 1993:72-73). After colonialism, the former colonial cities exploded in size and territory and developed complex internal structures, but the dependency of these cities did not fundamentally change. In this, there was a line of continuity with colonialism.

Major problems of the Third World metropolis

Rapid urban growth in Third World metropoles has been a catalyst for revolutionary social, cultural and economic transformations in both the cities and the countryside, with the so-called 'urbanisation of the countryside' and the 'ruralisation of the urban areas'.[14] Metropolitan growth has created numerous opportunities, economically, socially and culturally, and has to a great extent contributed to national economic growth. On the other hand, however, it has brought in its wake several serious problems such as overcrowding, infrastructural and economic bottlenecks, environmental degradation, unemployment and poverty, and inadequate housing facilities and basic social and urban services.

It is important to note that the problems that these Third World metropoles are facing today are not completely new, as if they have never been seen anywhere else in urban history before. Crowding, congestion, poverty and desperation were also found in the cities of Europe at the time of the Industrial Revolution and before. Moreover, the problems of the developing metropoles at present are not restricted to the Third World alone. A large number of Europe's cities are today also facing problems of congestion, unemployment, inadequate housing and transportation facilities, rising crime rates, increasing social and racial tension, and often high levels of pollution. What is new and different in the metropoles of the Third World, however, is the magnitude of most of these problems due to the dramatic growth and population increase since the 1950s.

Poor housing and widespread urban poverty are the most visible elements of rapid urban growth, and have received most attention in Third World urban literature and in the offices of governmental and non-governmental national and international aid agencies.[15] Figures put forward by Potter give an indication of the extent of the housing problem in the Third World. About half of all city dwellers in general are estimated to be living in housing which may be qualified as *sub-standard*, either in (authorised) slums, or in (unauthorised) squatter settlements (Potter 1992:24).[16] The ILO, in a comparable study of eight Third World metropolitan areas in 1988, estimated the slum population at between 12% in Seoul and 84% in Cairo.[17] Asia is generally considered the worst region in terms of housing, with national studies giving urban slum population figures of 54% for Indonesia, 47% for Bangladesh, and 36% for India (UN, 1996:7).[18] However, in recent years the housing situation in several African metropoles has become as bad as in many Asian cities.

So far as urban poverty is concerned, the latest report of the UN Population Fund once more indicates that 'It is still difficult to assess the nature and seriousness of urban poverty and suggest specific policy solutions' (UN 1996:5).[19] Urban poverty may, according to the UN, affect a third of all urban dwellers directly (the majority of them women), as they cannot adequately provide for their basic needs in shelter, employment, water, sanitation, health and education. However, its indirect effects are felt by the whole society, the UN argues, as the ability to meet the challenge of eradicating extreme poverty and providing basic needs will define and to some extent determine the viability of urban centres and the economies which they increasingly dominate. Estimates of the number of people living under the poverty line in Third World cities are not readily available and subject to almost insoluble problems of definition. One global estimate however, put forward by the UN, suggests a figure of approximately 28% for the entire Third World, naturally covering serious regional differences.[20] Nevertheless, national data often put the percentage of urbanites living in abject poverty at about 50%.[21] In general, what is certain, is that urban poverty has been increasing faster than rural poverty.

Both the housing problem and urban poverty are, however, a result of an underlying factor which was almost absent in the industrialising cities of 19th century Europe, i.e. the lack of sufficient absorptive capacity of the urban economy in relation to the increase in the number of potential job seekers. In contrast to European industrialisation, in the Third World of today urban development and growth are generally ahead of formal industrialisation, with relatively few jobs to be distributed amongst a large pool of potential workers. Consequently, most of these necessarily end-up employing themselves in jobs like street trading, shoe-shining, taxi driving, or in small cottage industries. This so-called 'self-help labour' in the informal sector has tremendously increased over the last couple of decades. Estimates of the size of the informal sector in the Third World in general are diverse and may be anywhere between 40 and 70% of the urban economy. Initially, in the 1950s and 1960s, the informal urban economy was considered a major problem and the ultimate goal was to integrate it into the formal economy. Later, however, the entirely different nature of both sectors was understood and their distinct contribution to the economy largely appreciated. Nevertheless, the lack of absorptive capacity in relation to the ever increasing number of job seekers remains, and unemployment and underemployment is still the single major economic problem in the Third World metropolis.[22]

It would be quite interesting to further discuss the relationship between urbanisation on the one hand and absorptive capacity and economic efficiency on the other. Cities doubling in size every decade would strain the capacity of most developed countries, which in the Third World has been occurring for several decades in an environment of low and strained human, physical and financial resources. To come to any generalising conclusions on the relationship between city size and economic efficiency is probably not possible, and will differ from city to city as it is influenced by many different variables. Nevertheless, much debate on this issue has been going on for several decades, and was related to the overall debate on urban development and planning strategies to be followed as to have the largest possible positive effect on national development. In the debate, the same polarisation between modernisation theory and its Marxist mirror of dependency theory became evident.

Urban development and urban planning strategies

Is further urbanisation and urban growth desirable, or has the Third World metropolis simply become too large and should further growth be prevented at all cost? In this section we shall argue that the continuation of the urbanisation process, even though it has produced a lot of undesirable side-effects, is nevertheless preferable, both for economic and social progress and for development. What is urgently required in most countries, however, is a well-planned urbanisation policy and proper urban planning policy, on a national level as well as on a local level.

National strategies and policies

Over the past fifty years, relatively few Third World governments have paid much attention to the way in which the urbanisation pattern was developing. The national development policies of the 1950s and 1960s, which were strongly influenced by western concepts of modernisation, were focused on industrialisation, and most of the private investments in industry were canalised to metropolitan areas.[23] However, many governments tended to focus their development efforts on only one or a few major cities, largely neglecting not only the rural areas, but also the intermediate cities and towns. Grave distortions and imbalances between urban and rural areas were the result.

In the 1970s and 1980s, many authors (Lipton, Todaro, Brown, Beier and others) criticised and blamed the Third World governments for this highly unbalanced, polarised development pattern, mainly because their policies were considered to be strongly biased in favour of urban development, and completely unsupportive to rural development. They therefore argued that the overall planning focus in these countries had to be drastically shifted from an urban bias (and overurbanisation) to a rural bias. The impact of this 'urban bias theory' on planning concepts was exceptionally strong. The majority of international aid agencies modeled their planning policies mainly after the urban bias viewpoint in academia, in favour of rural development.

A survey by the UN in 1978, on the geographical distribution of people in Third World countries, came to the conclusion that of the 116 developing countries surveyed, almost all of their governments considered the population distribution at that time to be 'highly unacceptable or unacceptable to some degree'.[24] Almost all believed that internal rural to urban migration was the major factor contributing to city growth. Therefore, most governments initiated policies to either slow down or reverse urban migration. Planning strategies were increasingly shifted towards rural development policies based on agricultural self-reliance, rural new town development, etc. and towards 'zero urban growth' and even 'deurbanisation' in some socialist Third World countries.[25]

However, these rural development and anti-urbanisation policies also received a lot of criticism, from those rejecting the existence of an 'urban bias theory', as well as from those recognising it and approving it.[26] Already in the 1970s the latter increasingly stressed that large metropoles were the economic engines of most Third World countries, and they feared that a drastic shift away from further urbanisation would in the long run have a devastating effect on overall economic and social development.[27] Cohen for example argued that large cities exist 'because of their competitive advantage for industrial production and the economies of scale associated with increasing urban size' (Cohen 1984:32). Furthermore, the proximity of suppliers and customers within the same urban area would lead to substantial savings in transportation, communications and trading costs. He therefore concluded that despite the number of disadvantages, large cities are bound to grow even further, and that there is no such thing as an 'optimum or maximum city size' for all cities.[28]

Considering the economic contribution of Third World metropolitan areas to their national economies, many of them producing over 30% of the GNP, it is indeed not desirable (if at all possible) to divert population growth away from the urban centres.[29] Average urban incomes are often more than three times higher than average rural incomes, and any policy aiming at bringing rural incomes to urban levels may prove unsuccessful. Even with a significant degree of rural development and a radical shift from urban bias to rural bias, the experience of the West indicates that urban migration will not significantly slow down, quite the contrary. Agricultural modernisation, as it was the case in all western countries, implies sooner or later an irreversible decline in the need for rural manpower. 'Developing countries', says the UN, 'cannot hope to modernise without a steady decline in the proportion of the working population engaged in food production' (UN 1984:110). In short, modernisation and urbanisation go hand in hand, and obstructing the processes of urbanisation and urban growth includes obstructing the modernisation process. The central question for all Third World countries should not be 'how to prevent cities from growing any further and how to keep more people out?', but rather 'how to manage the growing cities and how to absorb and to integrate the growing population?'.

One of the major problems of urban development in Third World countries in recent decades has been the absence of a creative strategy for urbanisation. Economic planning was rarely accompanied by urbanisation strategies, which is, given the fact that urbanisation is to a very large extent a result of economic planning, quite extraordinary. People like Harry Richardson (1984) or Bertrand Renaud (1981) strongly plead for the adoption and implementation of a 'national urban development strategy' (NUDS), especially for those developing countries that combine high rates of population growth with low levels of urbanisation, such as India or Pakistan. The goals of such a NUDS should be similar to those of other economic and social policies, and simple policies of population distribution are insufficient. They also consider an NUDS as complementary rather than an alternative to rural development strategies. Such a NUDS could take many forms, and must therefore be designed for each country individually, taking into account the specific characteristics such as urban history, population distribution and city system, the level of economic development, the political system, social structure, cultural environment and physical features (see also for example Costa *et al.*, 1988: chapter 1). Richardson suggests a large number of possible NUDSs, most of which involve attempts to somewhat slow down the growth of the primate city and to promote the expansion of a carefully selected set of secondary cities.[30] The major obstruction to such policies may be the extent of political acceptability and the courage and commitment of successive governments to implement them.

When it became clear by the 1980s that simply shifting development strategies away from the urban to the rural areas was not having the desired effects in terms of slowing down rural out-migration and uncontrolled metropolitan growth, the kind of ideas that Richardson and Renaud put forward, supporting further urban growth and development within the contours of a balanced NUDS, increasingly found acceptance with the international aid agencies and financial institutions, resulting in two important policy papers by the UN and the World Bank in the early 1990s.[31] At present, the central idea in these two papers is increasingly being implemented in a large number of Third World countries.[32]

Regional and local strategies and policies

Beside an overall national urbanisation policy, the countries and cities of
the Third World also need urban policies at the regional and local level.
Many different local strategies have been tried over the years, a few of
which have been reasonably successful, but most of them were put to
practice in isolation or complete absence of any national guidelines.
During the 1950s and 1960s most of these local policies were focusing
mainly on migration, as this was considered the single main cause of all
urban problems. Consequently, new incoming migrants had to be kept out
of the city, and therefore also any additional large-scale industry (as this
was considered the major pull factor). Key urban activities were
decentralised, there was disinvestment in social and infrastructural urban
services, and some local city authorities went so far as to introduce
residency permits, basically closing the city for outsiders.

   Most of these policies started from the questionable perception that
migration is the major direct cause of most urban problems, and it was
soon realised that the strategy to keep migrants out was not only
unsuccessful in itself, but also causing more harm than doing any good.[33]
Local urban policy should not directly focus on urban migration, but rather
on those areas that are the real problems in Third World metropoles: the
physical, financial, institutional, and managerial constraints.

   Physically, many current Third World metropoles are located, for
historical-colonial reasons, in low-lying coastal areas which are prone to
flooding and require large investments in drainage systems. Some cities
such as Bombay are located, for the same reasons, on an island or adjacent
to a mountain range, which adds for example large sums to the
construction and maintenance of a transportation network or to the possible
expansion of land area.

   Financially, metropolitan areas surely are not receiving too much
of the small pie of resources, as was claimed by the urban bias theorists.
Considering the number of people living in these cities, and the sometimes
enormous amount of tax money that is being collected here,[34] tax resources
have mostly to be shared disproportionately with province/state or nation,
and are (usually) relatively insignificant. Or as Sivaramakrishnan and
Green have put it, 'if a large city is to sustain its function as a principal
generator of wealth', it must be 'permitted to absorb a sufficient portion of

the wealth it creates for its own upkeep and further development' (Sivaramakrishnan and Green 1986:35).

In terms of institutional and managerial problems, most regional/local Third World governments in the 1950s and 1960s, were encouraged by the international aid agencies to shift their policies towards a project-oriented, sectoral approach. Cities increasingly started to undertake large-scale projects that were formulated, financed and managed sectorally by 'special public works authorities' (in housing, water supply, sanitation) with either regional or local jurisdiction. Although social and economic issues did not receive much attention, and although they tended to work independently from city councils or municipalities, these special public works authorities have been reasonably successful. Nevertheless, some problems remained, as the number of public corporations, statutory bodies, and parastatal agencies mounted rapidly, whereas the financial means to pay for their activities were not readily available, and agencies were competing for resources and opposing each other. Moreover, the undemocratic set-up of such agencies could be questioned, as the role of the appointed professional in city management increased at the expense of the elected representative.

Because of the problems between these agencies, since the mid-1970s more attention has been focused on the need for an additional, multi-sectoral response. The typical response in many countries was the establishment of 'development authorities' with metropolitan-wide jurisdictions. Initially, the tasks they were given were focused on intersectoral co-ordination and supervision, planning, programming, and budgeting. Increasingly, however, they have become involved in direct implementation of works and projects, which in many cases has led to a neglect of their prime multisectoral functions. Thus, ultimately, in most places where they have been established, these development authorities have increasingly taken over the responsibilities of sectoral agencies, and have failed to bridge the gap between the institutions responsible for creating assets and those responsible for maintaining and improving them. Moreover, at the same time they seem to have further undermined the tasks and functions of democratically elected municipal governments already weakened by the loss of sectoral responsibilities and financial and human resources to special *ad hoc* authorities. No wonder locally expressed needs at the municipal and neighbourhood level remain, in many instances, unheard.[35]

Once these financial, institutional and managerial constraints can be overcome, attention should be shifted to the two major problem areas in Third World metropoles, i.e. the insufficient absorptive capacity and housing.

Although it must be thoroughly understood that national policy decisions directly affect the ability of metropolitan areas to generate employment, the role of metropolitan management in promoting and creating employment may be the least understood of all its functions. Urban management has hardly addressed these basic economic issues, perhaps because employment promotion, over the years, has been considered primarily a task of the national authorities. Nevertheless, it is largely at the metropolitan level that the economies of agglomeration can be fully exploited and inefficiencies which adversely affect economic efficiency can be most successfully tackled. It is also at the metropolitan level that measures can be taken that directly support business and thus employment.

In terms of urban housing, many strategies have been developed and tried out, only some of which have proved to be useful. Initially it was believed that the government had an immense responsibility in providing housing for all, especially for the majority of urban poor. Later however, facing financial constraints and realising the complete failure of such programmes, these policies were almost everywhere abandoned and drastically changed. Since then the general housing philosophy was that 'where there is a housing shortage, the last thing Third World governments should do is to build houses' (Potter, 1992:35) and it was generally agreed that the role of government should be restricted to assisting the people in their efforts to help themselves. Since the early 1980s, the main international aid agencies strongly supported this vision, and three different forms of supported 'self-help housing' have developed: (a) the upgrading of existing slums and squatter settlements; (b) the provision of *sites and services* schemes; and (c) the provision of *core housing schemes*.[36] It is evident that the most important element of any of these low-cost housing schemes is first of all security of tenure, and secondly the affordability for the class of people for whom they are constructed.[37]

Conclusion

In this chapter we have attempted to provide an overall view on the urbanisation processes and urban growth patterns in the Third World in general, in theory as well as in practice. However, every country has its own specific past and its own specific economic, political and cultural characteristics, and such a general overview necessarily ignores numerous specific features, factors and elements that shape the urbanisation process and urban growth patterns in a country. In the next chapter we shall look into the processes and patterns of urbanisation and urban growth in India, home of the largest concentration of mega-cities in the world.

Notes

[1] Some examples of terms that are sometimes arbitrarily used: urbanisation, urban growth, urbanism ... city, metropolis, metropolitan area, conurbation, megacity, megalopolis... city system, urban system ... preindustrial city, post-industrial city, colonial city, western city, socialist city ... new town, countermagnet, growth area ... etc.

[2] In order of size (>) these eight will be Bombay, Karachi, Dhaka, Calcutta, Delhi, Lahore, Hyderabad, and Madras.

[3] As the UN Population Fund observes, migration is very difficult to study, since a complete analysis requires information on people and conditions in at least two different places and at various times. Information about communities, families and individuals is needed.

[4] Angotti puts the number of people that moved from current Bangladesh to India (mainly to West Bengal and to Calcutta) within the five to six years after Independence and partition at roughly 3.5 million (Angotti 1993:87).

[5] According to Todaro (1984) urban natural increase was for 60.7% responsible for urban growth in the Third World in general. According to Preston, urban growth percentage in the largest Third World countries due to natural increase was 67.7% in India (1961-71), 64.3% in Indonesia (1961-71) and 55.1% in Brazil (1960-70). According to the UN, the percentage of natural increase in urban growth has increased between 1960 and 1980 from around 58% to over 70% in Africa, from 59% to 66% in Latin America and has decreased from 59% to slightly more than 50% in Asia (excluding China) (UN 1996:27).

[6] Some of these points were also raised by Kingsley Davis (1984) who already in the mid-1970s wondered where the change in agriculture would come from that

would enable the huge projected urban population to subsist, as there would be ever more farmers, not fewer, to consume the agricultural produce.

[7] Potter (1992) provides some World Bank data (1972) for the decade 1960-70, with the then still relatively small African cities growing most rapidly: Abidjan 11.0%, Kinshasa 10.0%, Kampala 9.2%, Dar es Salaam 9.0%, Seoul 8.0%, Katmandu 7.9%, Bogota 7.0%. The already large urban areas such as Mexico City, Cairo, Bombay or Calcutta were during this decade all growing at less than 5% p.a.

[8] As Renaud observes, a country can, however, experience primacy without having any very large city, while another country can have very large cities without exhibiting such primacy (Renaud 1981:32). For reflections on primacy in Asia see for example Norton Ginsberg's contribution in Costa *et al.* (1988).

[9] Sjoberg and others have classified cities as 'pre-industrial' and 'industrial', according to the socio-economic characteristics of the cities (see Sjoberg 1960, *The Pre-Industrial City, Past and Present*). Naturally, it didn't take long before some more recent theories started describing modern cities as *post-industrial cities* and the societies as *post-urban societies*. According to Kopardekar, this type of classification actually obscures the nature and role of the process of urbanisation and that of city growth in the process of development (Kopardekar 1986:41).

[10] Many theories and works on urbanisation concentrate on some selected aspects of the process of urbanisation, without taking into account the larger perspective. Moreover, some works have been looking at urban growth trends or city growth in Third World countries, whereas others (McGee, Breese, Hauser) have tried to explain growth trends.

[11] Manuel Castells paid a lot of attention to the primacy of urban politics, which according to him was the major explanatory variable of the characteristics of urbanisation (Castells 1983:194).

[12] As usual, many different views on urbanisation and its relation to dependency theory exist within Marxist writing. A general point of critique is the tendency, due to the large output of Marxist writings in Latin America, to over-generalise about Latin America as a whole. Another critique is that a large number of studies on urbanisation were reduced to debates on urban economic and social structures, interpreted in isolation from the urban-rural debate. Further, it must be noted that many urban specialists almost exclusively concentrated on the capitalist economic structure, which seems to be a fairly narrow-minded approach of the topic.

[13] Sivaramakrishnan and Green have stressed that the term *metropolis* has produced different perceptions according to the field of study in which it is being used. They point to the ancient Greeks, to whom the metropolis had the meaning of the mother city and principal seat of government of a state or colony. Urban geographers and planners now tend to use the term to refer to a large identifiable area of continuous urbanisation consisting of several administrative jurisdictions.

Demographers today often classify cities with populations of more than 1 million people as metropolitan, and in common usage the term is widely employed to symbolize social, economic, and political status.

[14] In the countryside, the rapid urban growth process had a large impact on sex ratio, fertility, labour organisation and distribution, income, consumption pattern, size of agricultural land, family structures, and traditional habits and attitudes. In the urban areas, stability has been replaced by explosive changes at a socio-cultural and economic level (from the large or joint family system to the small family system, in religious practices, in social stratification and polarisation, in dress and food habits, in economic organisation, production and consumption patterns, and in the general way of living), as well as at a physical and functional level, as the rising flood of people quickly eroded old city boundaries and overwhelmed health, sanitary, environmental and educational facilities and services.

[15] The first worldwide conference on housing problems in the Third World (*Habitat I*) was held in Vancouver in 1976. Due to a narrow-minded focus on the housing problem, the conference had limited success. In 1987 it was the *International Year of Homeless People*, when Third World housing, again, attracted a lot of international attention. Twenty years after *Habitat I*, the world's leaders thought that it was time for *Habitat II*, which was organised in Istanbul in May 1996. All of these large world housing conferences are characterised by well-meant but completely unrealistic goals, such as *shelter for all by the year 2000*, mentioned in about every report during the International Year and in those of Habitat II.

[16] Grimes, writing on the subject, indicates that analysis of data on housing must be used with great caution. Reasonably accurate data are often lacking, and terms such as *slum* or *squatter* are pejorative rather than descriptive (Grimes 1984:168).

[17] Figures for some other cities were 58% in Lagos, 57% in Bombay, and 25% in Mexico City. The high figure for Cairo is probably due to a different definition used by the Egyptian authorities of what a *slum* is (including for example all informal housing, whatever their physical condition).

[18] Some argue that the wide attention for housing in Third World cities has been exaggerated and that the significance of housing is much more limited than is generally presumed. They point to the fact that housing often does not have high priority among squatters, who consider economic conditions more important than housing, and to the fact that housing conditions in the rural areas are worse than in the metropoles (see for example Nientied and van der Linden 1988:145-46).

[19] The UN agency further adds that 'conditions of life in the cities, especially for the most vulnerable groups, are either poorly documented or the data are difficult to access, and are underanalysed and underused'.

[20] The figure for the urban population of sub-Saharan Africa is estimated at 41.6%, for Latin America at 26.5% and for Asia at a relatively modest 23%.

[21] As the UN suggests, even these figures may be an underestimate, as official poverty lines are often set unrealistically low, below the levels required to meet basic needs.

[22] Underemployment is the widespread phenomenon that many are doing a job that could be done by far fewer. Because of such underemployment, the general employment problem in Third World cities is further underestimated.

[23] Investment in heavy industry was, as in India, often outside metropolitan areas.

[24] UN Economic and Social Council (Population Commission, 20th Session), *Concise Report on Monitoring of Population Policies*, Dec. 1978:27-28.

[25] We shall come back to the 'new town concept' in chapter 5.

[26] In this period a large debate evolved between supporters and opponents of the *urban bias theory* in development policy, but for reasons of limited space we cannot enter this discussion.

[27] Already in 1967 it was in fact agreed, at the *Pacific Conference of Urban Growth* held in Honolulu, that the cities and urbanisation in general were the engines of growth in the Third World. Two years later, at another conference in Hong Kong, it was agreed that cities, despite their problems, were here to stay and needed intensive research, and the relationship between urbanisation and development was seen as a positive one (for more details see Ginsberg 1989:19-23).

[28] Both the debates over scale economies and the possible diseconomies in some services, and over the optimal city size at which scale economies are optimised, remain largely unresolved.

[29] In the mid-1980s, almost 50% of Sri Lanka's employment in commercial, financial and transport services was accounted for by Colombo alone; over 60% of the Philippines' manufacturing plants were located in its capital Manila; and one-third of Thailand's GNP originated in Bangkok (Sivaramakrishnan and Green, 1986:34). About 60% of the total GNP of all Third World countries put together is produced in the cities (Internationale Samenwerking, 1996).

[30] This strategy to divert part of future urban growth from the very large to the medium-sized and smaller urban settlements is currently being followed in several countries to tackle the major problems of excessive metropolitan growth.

[31] *Urban Policy and Economic Development: an Agenda for the 1990s* (World Bank Policy Paper 1991) and *Cities, People and Poverty: Urban Development Co-operation for the 1990s* (UNDP Strategy Paper 1991). The ideas in these papers have been discussed, and in general agreed, at a large workshop with representatives of international aid agencies and Third World governments in

November 1991 in London (organised by the DPU, University College London and the ODA).

[32] It nevertheless took the World Bank quite some time to realise the importance and growing impact of urbanisation processes on Third World development. Only in 1972 it published its first working paper on urbanisation, which reviewed existing material on the subject done by other agencies such as the UNDP and by individual scholars. In the paper, the World Bank argued in favour of a selective project approach and a learning-by-doing approach. This led to an expansion of Bank assistance to Third World governments (World Bank, June 1972, *Urbanization*; Working Paper, Washington DC).

[33] Some sectors of the economy that are based on seasonal labour suffered, and corruption rapidly increased.

[34] In the mid-1980s, Bombay for example accounted for nearly 80% of all tax revenues in Maharashtra state, and for about 25% of the entire country's income tax receipt (Sivaramakrishnan and Green 1986:34).

[35] All this however does not mean that the concept of development authorities, covering the entire metropolitan area, was a bad idea and is useless. On the contrary, what may be needed to make it work is a matching intermediate political level, between the traditional national or regional level on the one hand, and the local, municipal level on the other. Today, very few Third World countries have such a responsible government at a metropolitan level. For a strong argument in favour of this strategy, see for example Angotti (1993).

[36] The successful implementation of slum and squatter upgrading schemes is largely dependent on the provision of basic services and the improvement of the physical lay-out (besides the security of tenure); in sites and services schemes small plots of land (mostly newly developed land) are serviced with access roads, water and electricity, and people are left to decide for themselves how and when they build their house. To be successful, nearby employment opportunities are needed (beside security of tenure); Core housing schemes are quite similar to sites and services, with that difference that on the reserved plots of land core units have been provided, sometimes comprising just the floor and a toilet, sometimes also brick walls, a roof, a kitchen, and so on.

[37] When we will discuss the implementation and success of such schemes in New Bombay, we shall return to the self-help topic.

3 Urban Development in India

Introduction

As Ginsberg pointed out in a paper of 1989, even within one region of the developing world such as Asia 'the differences in urban types and systems from country to country, despite great similarities, are striking' (Ginsberg 1989:21).

Although the broad lines of the urban development process in Asian countries may look quite similar, their political, economic and cultural histories have been very distinct, and with that also their urbanisation processes. Hence the necessity to look at the urban development process in India separately.

Academic interest in the urban growth and urbanisation process in India (and in Asia as a whole) is a recent phenomenon and has only gained momentum in the post-Independence period. Although some scholars such as Patrick Geddes worked individually on Indian cities as early as 1915, their work was largely descriptive in nature and based on secondary resource material (mainly archives). From the 1950s the focus somewhat shifted to a more analytical approach and primary sources based on personal fieldwork were increasingly being used, for instance in the preparation of urban plans for metropolitan areas. At present, some fields within the social sciences (geography, town planning) have established a fairly strong reputation in urban research in India, whereas in other fields (politics for instance) the interest in urban studies is still rather limited. Overall, however, the academic literature on urban development in India has become quite diverse and has grown into a mature subdiscipline within the social sciences.

Looking at the urbanisation process in India is not an easy task. Simultaneously, we need to look at a period spanning nearly 5,000 years, at a huge area the size of the 15-nation European Union, at people from widely different ethnic origins with a widely different cultural background, and at a complex economic-political framework. It is mainly this framework that has shaped the urban development process. Moreover, the fact that so many social sciences are currently involved in the study of certain aspects of urban development indicates the strong multidisciplinary character of the topic. Therefore, this account of the urban development process in India will be very brief and incomplete, and will only touch upon the most important aspects of this process.

In a first section of this chapter we will discuss the broad historical background of the urban development process on the Indian subcontinent. In a second section we will briefly look at the modern processes and patterns of urbanisation. In the third section of this chapter we will look at Indian urbanisation policies and urban planning concepts since Independence.

Urban development in India in historical perspective

The pre-colonial period

The Indian subcontinent has one of the most ancient urban traditions in the world. The Harappan civilisation, along the Indus river in current Pakistan, dates back to well before 2,500 BC and flourished over a period of at least 600 years. Some of the cities of this civilisation, the most famous of which were Mohenjo Daro and Harappa, had a population of over 30,000, showed a fairly complex and advanced culture, and despite their wide area distribution they also showed strikingly similar urban patterns and building concepts. They are therefore widely recognised as the first examples of real town planning.[1] By about 1,700 BC the Harappan civilisation had come to an end for reasons which are today still not entirely clear.[2]

For about a thousand years after the Harappan civilisation there is no proof at all of any urban development in the area. Only by the 6th century BC there was again a political organisation strong enough to sustain an urban civilisation, associated with the two closely related cultural streams of India, the Aryan civilisation in the north and the Dravidian civilisation in the south. From then, India has had a more or less continuous history of urbanisation for about 2,500 years, obviously with both periods of urban growth and periods of urban decline.

In North India, from the 3rd century BC a major part of the Indian subcontinent came with the Mauryan emperor Ashoka under one political authority, and both the Indus and Ganga river basins saw the development of numerous towns and cities. At around 180 BC the Mauryan empire collapsed, although cities kept on growing even in the post-Mauryan period (180 BC-600 AD). During the post-Gupta period (roughly from 600 to

1,000 AD) cities in the northern region declined and were largely neglected.

At the time of the collapse of the Mauryan empire in the north, towns also started to develop in the southern area under the Tamil kingdoms. As in the north, these southern towns were not just centres of defence, administration and economic activity, but were also centres of culture and religion. Contrary to the evolution in the north, however, cities and towns in the south did not collapse, but continued to flourish between 800 and 1,200 AD.

With the arrival of the Muslims, from the 9th century onwards, urban development in North India was revitalised, and by the time that the Muslim rule was firmly established with the foundation of the Mughal dynasty in the 16th century, the urban pattern in the north was transformed. Because of the centralisation of political control and power in the cities, the cities became the centres of power and wealth. And because of the import of an alien culture they also became the focus of new architectural styles and new patterns of urban living, to replace those that had been largely destroyed. Great mosques, mausoleums and massive fortresses were built and impressive new cities such as Agra, Delhi and Fatehpur Sikri were created. During the same period, the cities and towns in the south largely remained out of reach of the Mughals, and the temple and port towns prospered mainly on the basis of local industries and trade with the West.

In general, urban development in the pre-colonial era, spanning a period of roughly 3,500 years, rarely generated towns or cities with a population exceeding 100,000 people. Nevertheless, the broad regional pattern of an urban hierarchy, with the largest concentration of towns in the northern belt of the subcontinent and a significant number of important cities in the small kingdoms of the south, had by the end of this period clearly been laid out. Central India, due to its relative inaccessibility, remained sparsely populated (Bradnock 1989).

The colonial period

The British came to India at a time when the subcontinent was 'perhaps the most urbanized nation in the world' (Ramachandran 1989:22). However, during the first phase of British rule the level of urbanisation in India declined. Soon afterwards it would pick up again and would ultimately become higher than ever before.

The British traders whom in 1639 first settled in the south on undeveloped land at current Chennai (Madras), established two more trading posts; one on the west coast (Bombay in 1668), and one on the east coast (Calcutta in 1690). Although these new trading settlements developed fairly slowly at first, their rapid growth during the 19th century would cause the greatest urban transformation in Indian history. These major port cities grew mainly because of trade and commerce. The restrictions to rapid urban growth that were prevailing in earlier times were slowly removed by the construction of an extensive rail and road network that embraced virtually the entire subcontinent, and of which these cities were the major foci. In the meantime, India also slowly developed its own industrial base, with these port cities becoming the main centres of manufacturing.

Calcutta, Bombay and Madras thus became not just the main seats of colonial power, but also the main centres of economic change. The overall effect was that the Indian urban landscape was, as Murphey observes, 'no longer dominated by inward-centred administrative or religious foci or local-regional markets, but by the burgeoning ports that faced outward to the world overseas' (Murphey 1996:23).

This new urban pattern, which was emerging by the late 19th century, would be consolidated during the first half of the 20th century. Due to the economic importance of Bombay, Calcutta and Madras, and increasingly also of Delhi after it became the new political capital of British India in 1912, these large cities became a major attraction pole for hundreds of thousands of textile workers and other workers in the manufacturing industries.[3] Later, in the post-colonial era, tertiary sector activities would take over the prime economic role that the manufacturing industries had been playing in these cities during the colonial period.

Modern processes and patterns of Indian urbanisation

Population growth, urban growth and urbanisation in the 20th century

Whereas during the 20th century the growth of the Indian population never exceeded 2.5% per annum, the absolute increase in population has been quite impressive. In 1901, India already had a total population of 233 million. In the early phase of Independence (1951), that number gradually increased to about 361 million; a growth rate of less than one percent per year on average.[4]

Table 3.1: Urban versus total population in India, 1901-2025

Year	Total population (millions)	Av. growth per annum	Urban population (millions)	Av. growth per annum	% Urban
1901	233.0 (a)		25.6		10.99
1911	252.1	0.57%	25.6	0.00	10.15
1921	251.3	- 0.04%	27.7	0.67	11.02
1931	279.0	1.10%	33.0	1.75	11.83
1941	318.7	1.42%	43.6	2.79	13.68
1951	361.1	1.33%	61.6	3.46	17.06
1961	439.2	2.16%	77.7	2.31	17.69
1971	548.2	2.48%	107.0	3.21	19.52
1981	683.3	2.47%	156.2	3.78	22.86
1991	846.3	2.38%	217.6	3.64	25.71
2001	*1016.0 (b)*	*2.00%*	*307.0 (c)*	*3.22*	*30.20*
2025	*1392.1(d)*				

Source: Statistical Outline 1994-95 (p. 42-3, 48). Notes: (a) population figure 1901 from Mills & Becker; (b) personal projection on basis of modest average growth rate of 2.0% per annum; (c) projection Statistical Outline 1994-95 (p.42); (d) projection UN 1996; urban population figures from Mills and Becker (1986) except 1991 figure (Statistical Outline).

Figure 3.1: Urban population vs. total population in India, 1911-1991

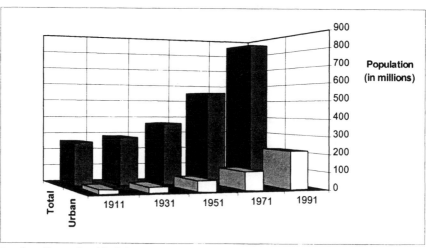

Based on figures in table 3.1

Since then, however, the growth rate increased by 2-2.5% per year, and the absolute number of people rapidly increased to 846 million by 1991. This is more than the total population of Europe including Eastern Europe (727 million) and more than the total population living on the African continent (748 million) or on the American continent (785.7 million). Even with a relatively modest growth rate of 2.0% per annum for the decade 1991-2001 India will cross the one billion population mark by the year 2000, and the projection for 2025 is nearly 1.4 billion (UN 1996).[5]

Even more dramatic was India's increase in urban population. Although there has not been a clear overall pattern of urban growth over the entire period,[6] the overall urban growth rate since 1901 was with an average of 2.4% per annum relatively modest. If this period is split into two, the average urban growth per annum was only 1.3% between 1901 and 1941, but nearly 3.3% between 1941 and 1991.

The overall level of urbanisation in India has always been rather low and fairly constant. From as early as the 6th century BC to about 1900 AD, the level of urbanisation was probably never lower than 5% and never higher than 12% (Ramachandran 1989:91). Since the beginning of this century there has been a slow but steady upward trend in the level of urbanisation. Between 1901 and 1931, urbanisation figures were within the narrow margin of 11 to 12%, but since the 1930s the proportion of urban population increased to 17.0% in 1951, 19.5% in 1971 and 25.7% in 1991. According to projections, the Census of 2001 should register an urbanisation rate of just over 30%. Thus, compared with other countries, in the western world as well as in the developing world, India's urbanisation rate can still be considered as fairly modest.

Nowhere, however, is the distinction between urbanisation and urban growth more crucial than in India. Whereas the proportion of the Indian population living in towns and cities may be rather low, the absolute increase in urban population, especially in recent decades, has been tremendous. As recently as 1951, the total urban population was 'only' 61.6 million. Twenty years later, in 1971, that number had increased to 107 million, and during the following twenty years the urban population more than doubled to 218 million. The number of 'urbanites' in India for 1996 has been put at 257 million (UN 1996) and the projection for 2001 is 307 million (Statistical Outline 1994-95). With that, India's current urban population matches the total population of the US, and is larger than the total population of any other country in the world apart from China.

Factors of urban growth

As we have seen in the previous chapter, urban growth is the sum of urban natural increase plus rural to urban migration plus city limit extensions. To make any broad statements on the balance between urban migration and urban natural increase as contributors to urban growth in India could be dangerous. Detailed information on urban migration has never been provided by the censuses, and reliable data are lacking. The main reason for this may be that urban growth, natural increase and urbanisation and migration, although they may seem to be fairly simple processes, are in fact very complex. They constantly influence each other, and are at the same time influenced by many different external variables themselves. This makes a correct collection and interpretation of data over a given time period extremely difficult.

Some authors have calculated the contribution of urban migration to urban growth in an indirect way. For example, rates of natural increase in rural and urban areas have for a certain period been compared with each other, and when it was found that these rates were almost identical (as they were in India during the 1970s for instance), it was presumed that any increase in the rate of urbanisation (as the proportion of urban to total population) had to be entirely the result of urban migration (Bradnock 1989). Based on this approach, Crook and Dyson have calculated that in the 1970s about half of the absolute urban growth in India must have been the result of migration, a percentage which according to them had been steadily increasing over the preceding decades (Ibid.). However, for the decade 1961-1971, Ramachandran calculated that no less than 24 million of the total absolute urban increase of about 29 million people was due to rural-urban migration. General and reliable figures for the most recent decade 1981-91 are not readily available, not for India as a whole, nor as Ramachandran observed for cities individually (Ramachandran 1989:14). For the late 1980s, Ramachandran has put the absolute figure of rural to urban migrants in India per annum at 'about 3 million' (Ibid.:91). However, the impression that is given by the recent experience of some of the metropolitan cities (that have more reliable figures), is that the contribution of urban migration to urban growth has today become less important than natural increase. Of course, this does not necessarily indicate that urban migration in absolute numbers is slowing down, it simply points to urban populations which by now have become so large that their natural increase figures exceed the urban migration figures.

The urban migration process is reflected by the urban sex ratio, which in the four mega-cities is between 827 and 831 females per 1000 males (Statistical Outline 1994-95:209), and by the shape of the population pyramids of most large cities. The largest population group are young males in the age group of 20 to 39 years old. They migrate to the city, some on their own and others with their family, some permanent and others temporarily, in search of a job.[7] The main reasons for urban migration in India are, therefore, economically induced. Poverty, which may be the result of many different factors, is clearly the most important push factor in this process, whereas the job opportunities are the major pull factors. Nevertheless, these are crude generalisations which need to be interpreted as such.[8]

Apart from economically induced urban migration, cultural and political factors or natural calamities may also instigate rural people to migrate. Education, public health facilities, an escape from the cultural and socio-economic barriers inherent in the caste system, the tradition for females to move into the house of the husband's family, and the general excitement of city life are all possible socio-cultural factors that may influence a decision to move out. In political terms, important events like the 1947 partition or large-scale public infrastructure projects have also, at regular intervals, played an important role in the urban migration process. Finally, regular natural calamities add to the urban migration process in India. In most cases the real causes for urban migration are probably a combination of different factors, and in general it could be argued that potential migrants move simply 'because they believe the advantages of their place of destination outweigh those of the place they are currently living in' (Bradnock 1989:15).

The impact of the third factor in the urban growth process, the geographical extension of city boundaries, is often underestimated and sometimes even neglected. The Census of 1981 recorded an increase of at least 10,000 km² of urban area, due to city limit extensions and the transformation of rural land into urban land (Amitabh 1994:39). This territorial extension is much more pronounced in northern India than in southern India. The expectation of the NIUA is that by 2002 another 13,000 km² of rural land will be incorporated as *urban*. The impact of such incorporations may be quite large. For example, the total urban growth of class III to class VI cities (see below) given in the Census of 1971, is believed to be almost entirely the result of such incorporations (Ibid.).[9]

Structure and pattern of urbanisation

Since urban population growth differs from city to city, and since urbanisation occurs unevenly over space, we also need to have a brief look at the structure and pattern of urbanisation. The Indian censuses have always collected data on the basis of six different city size classes.

Table 3.2: City size classes in India

Class I	100,000 or more (these places called *cities*)
Class II	50,000 - 99,999 (these and smaller places called *towns*)
Class III	20,000 - 49,999
Class IV	10,000 - 19,999
Class V	5,000 - 9,999
Class VI	5,000 or less

Although Mills and Becker have argued that India's definition of *urban* is quite acceptable, significant variations in the application of the census definition have occurred from state to state, and have caused confusion. In 1981 for example, Uttar Pradesh counted 659 census towns compared to less than half (325) in the previous decade, whereas Tamil Nadu counted 245 towns in 1981, compared to 400 ten years earlier.

As can be seen from table 3.3 and figure 3.2, until 1991 India's urbanisation pattern showed a tendency for the urban population to concentrate in the larger (class I) cities. Whereas in 1951 the large cities (100,000 or more) accounted for nearly 45% of the total urban population (coming from 26% in 1901), by 1991 they already accounted for more than 65%. These cities, nevertheless, constitute only 8.2% of the total number of towns and cities in the country. The medium-sized cities and towns (class II and III - 20,000 to 100,000) accounted for nearly 26% of the total urban population in 1951 (coming from 27% in 1901), a proportion which has remained more or less stagnant over the decades until 1991. The small towns (class IV, V and VI - 20,000 or less) accounted for nearly 30% of the total urban population in 1951 (coming from 47% in 1901), but by 1991 their share had sharply decreased to less than 11%. These towns, nevertheless, make over 56% of the total number of towns and cities in the country. Grouped into three categories this gives the following pattern:

Table 3.3: Population distribution (%) over city size classes, 1901-1991
(absolute figures between brackets, in millions)

	1901 (25.9)	1951 (62.4)	1961 (78.9)	1971 (109.1)	1981 (159.5)	1991 (217.6)	1996 (257.3
Class I	26.0 (6.73)	44.6 (27.83)	51.4 (40.55)	57.2 (62.35)	60.4 (96.34)	65.2 (141.9)	
Class II	11.3 (2.93)	10.0 (6.24)	11.2 (8.84)	10.9 (11.88)	11.6 (18.50)	10.9 (23.72)	
Class III	15.6 (4.04)	15.7 (9.80)	16.9 (13.33)	16.0 (17.44)	14.3 (22.81)	13.2 (28.72)	
Class IV	20.8 (5.39)	13.6 (8.49)	12.8 (10.10)	10.9 (11.88)	9.5 (15.15)	7.8 (16.97)	
Class V	20.1 (5.21)	13.0 (8.11)	6.9 (5.44)	4.5 (4.91)	3.6 (5.74)	2.6 (5.66)	
Class VI	6.1 (1.58)	3.1 (1.93)	0.8 (0.63)	0.4 (0.44)	0.5 (0.80)	0.3 (0.65)	
Total	100	100	100	100	100	100	

Source: Statistical Outline of India 1994-95; 1996 figure from UN (1996)
Note: absolute figures my calculations

Within Class I, the larger cities have attracted the largest share of the urban population. Between 1961 and 1991 the number of cities with a minimum population of 500,000 has increased by about five times, from 11 to 53 (Nath 1989:31 and Census 1991). Furthermore, the number of metropolitan cities (1 million +) has increased decade after decade, from only 5 in 1951 to 23 in 1991 (Amitabh 1994). Between 1991 and 2000, the official projection is that India will add another 13 metropolitan cities to this list.

Figure 3.2: Distribution of urban population over city size classes (1901-1991)

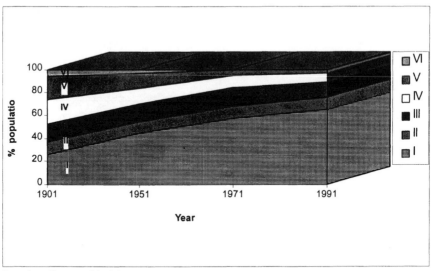

Based on figures table 3.3

**Figure 3.3: Population distribution over large, medium, small
cities/towns 1901-1991**

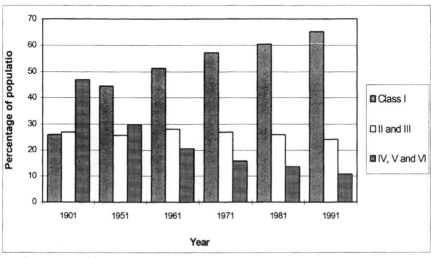

Based on figures table 3.3

Table 3.4: Number of 100,000, 500,000 and 1 million cities 1951-2001

	100,000 +	500,000 +	1,000,000 +
1951	---	11	5
1961	---	---	6
1971	---	---	9
1981	216	---	12
1991	296	53	23
2000	---	---	36

Source: Mohan 1996, Nath 1989, Amitabh 1994, National Commission on Urbanisation '87

In 1991, the 23 metropolitan cities housed about one third (72 million) of the total urban population of India. More than half (52%) of this population of 72 million lived in the four largest cities (Bombay, Calcutta, Delhi and Madras) (Statistical Outline 1994-95:pp.50-1). Compared to 1981, their share in the total population of the (then only 12) million-cities has decreased (coming from 65%). The growth rate in these four largest cities has decreased since about 1970, to less than 50% between 1981-91.

Although the larger cities have grown much faster than the smaller cities and towns in absolute figures, the small towns show a faster growth rate than the large cities. Moreover, the growth rate of the metropolitan cities is now less than that of the larger (100,000 to 1 million) cities (Datta 1992:194).

India's urbanisation pattern also shows significant differences in levels of urbanisation from state to state. If on the basis of the 1991 Census the states are categorised into three groups, four states (Maharashtra, Tamil Nadu, Gujarat and Karnataka) come out (in this order) as 'highly urbanised' states (30% or more urban, with Maharashtra -38.7%- the most urbanised of all). Four other states (Uttar Pradesh, Orissa, Bihar and Assam) come out as 'lowly urbanised' states (less than 20% urban, with Assam -11.2%- the least urbanised of all). All other states are 'medium urbanised' (20 to 30% urban) (Statistical Outline 1994-95:pp. 40-49).[10] In the previous census of 1981 three states were classified as 'highly urbanised' (i.e. Maharashtra, TN and Gujarat), and five states as 'lowly urbanised' (Ibid.). In general, western and southern India are thus relatively highly urbanised, while eastern and northern India are relatively lowly urbanised.

Causes of urbanisation

Numerous data world-wide indicate the close relationship between urban growth and economic development. As we have seen earlier, the major spurt in urbanisation in Europe and the US coincided with the emergence of the Industrial Revolution. In a similar fashion, the urbanisation process in India has also closely followed the process of economic development. As elsewhere, private industries in India, so far as they were not restricted by government intervention, have always tried to exploit the economies of agglomeration and have favoured the larger cities and towns for their business location. Consequently, since the 1950s, India has seen an enormous concentration of economic activity in and around its four major cities (Bombay, Calcutta, Delhi and Madras) and to a lesser extent in and around the other million-cities. Ramachandran has calculated that during the 1980s the four major cities alone contributed more than 70% of the income tax revenue at the national level (Ramachandran 1989:323).

Of course, India is not a unique case in this respect. Always and everywhere cities have functioned as the major engines of economic growth. It is, then, no coincidence that in 1991 the lowest levels of urbanisation were found in the economically weakest states of Bihar, Orissa and Assam, and that the highest levels of urbanisation were found in the economically most advanced states of Maharashtra, Tamil Nadu and Gujarat.

Strong urban economies are not the only prerequisite for large-scale rural to urban migration however. The experience world-wide is that modernisation in agricultural production is as necessary.[11] An increasing surplus of rural labour, due to modernisation, will automatically lead to rural out-migration, and will search for alternative occupations in the urban industrial sector. In the meantime, the latter attracts and creates new activities in the services sector, which in its turn becomes an important new field for alternative employment.

This pattern from mainly primary to predominantly secondary and tertiary activities is a pattern that every economy and country will experience at some point. The difference in present-day India, compared to the experience in western countries and in numerous developing economies, is that the industrialisation process and the increasing activities in the tertiary sector do not yet coincide with a large-scale modernisation process in agriculture. Consequently, a significant decrease in rural employment is in India only slowly emerging. Hence a continuing high rural population growth and a relatively low urbanisation rate.

In contemporary India, however, one has to be aware of a clear distinction between production and employment. Whereas in 1951 the share of agriculture in the Indian GDP was still 60%, by 1992 it had decreased by nearly half to only 32%. In the meantime, the share of both industry and services had significantly increased to 27% and 40% respectively by 1992 (Statistical Outline 1994-95:216 and Amitabh 1994).

In terms of employment, on the other hand, the share of the industrial sector in total employment had decreased to about 11% and that of the services sector had increased to 27%. In the meantime, the share of the agricultural sector, despite its decreasing share in GDP, was still by far the highest (62% of the total labour force in 1994, coming from 73% in 1973) (The World Guide 1997/98:299). The statement that India is still predominantly rural is thus only partly correct.

Figure 3.4: Production and employment per sector (1992/1994)

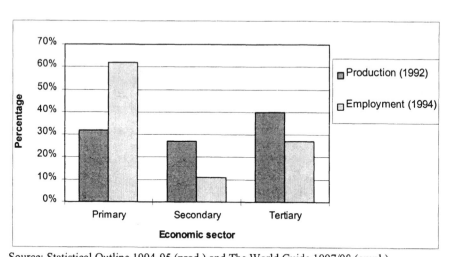

Source: Statistical Outline 1994-95 (prod.) and The World Guide 1997/98 (empl.)

These economic indicators and nuances are of course quite important in terms of future predictions of urban growth patterns in India. Given the economic liberalisation policies of the different Indian governments since 1991 (the Congress Party, the coalition of regional and leftist parties, and the coalition formed around the BJP), serious economic growth, predominantly in the industrial and services sectors, may be expected in the coming decades. If this happens, and the modernisation process in agricultural production continues and accelerates, it is inevitable that the urbanisation rate in India will rapidly increase and that the cities will continue to grow at a very fast pace.

Urbanisation policies and urban planning

Confronted with a highly unbalanced urbanisation pattern and a very rapid urban population increase, highly concentrated in the largest cities, the Government of India, over the years, has developed a number of *urbanisation policies* in an attempt to diversify such city-based population concentrations. Simultaneously, at a regional level, state and city authorities have tried to implement a number of *urban policies* in an attempt to relieve the largest cities of the increasing pressures of a constantly growing population and deteriorating urban congestion. However, so far as implementation is concerned, both urbanisation policies and urban policies can at best be called moderately successful.

Urbanisation policies

While India has a tradition of detailed national planning (industrial planning, agricultural planning), curiously enough the country never developed an explicit national urbanisation policy as such. This does not mean however that there are no policy guidelines in this field at all. They are scattered around and can be found mainly in laws, government reports, and in the national Five Year Plans.[12]

Since Independence, Indian policy makers always seem to have been convinced of the correctness of the Gandhian anti-urban philosophy and of the existence of 'overurbanisation' (which, as N.V. Sovani has forcefully argued, is a myth). The same policy makers have also considered urbanisation and industrialisation to be almost synonymous (an assumption which is of course not correct either). Consequently, Indian policies on urbanisation have been rather limitative and have in practice been restricted largely to industrial licensing and to incentives for decentralised industrial investment. In 1951, a license for all new or expanding industries was made compulsory for the first time. Five years later the 'Industrial Policy Resolution' was passed, which aimed at using licensing as a means to achieve balanced regional development. Since then, various industrial licensing policies were passed in order to restrict further growth in the largest cities: government-owned industries were preferably established in rural areas or smaller towns (and were sometimes relatively successful); small-scale industries received full government support; large industrial estates were established to encourage such small-scale industries and to disperse further industrial growth in and around the largest cities; licensing to new or expanding industrial corporations within the boundaries of metropolitan cities was prohibited since 1977; and certain advantages (infrastructure facilities, tax benefits,...) were given to large-scale industries setting up businesses in backward districts.[13]

Whereas the extent to which such industrial licensing programs has been successful in terms of industrial dispersal is debatable, it is most doubtful whether they have had any effect on the growth of large cities. Policies which are unilaterally focused on development of smaller towns and rural areas will never be able to stop further growth in the metropolitan cities. Such policies may at best somewhat slow the growth of these cities. Moreover, the desirability of such policies and programs must, again, be questioned. Mills and Becker were quite right when they warned that 'misguided attempts to reduce the size or growth rates of large cities can do great harm', and they concluded that the greatest danger of such policies is that 'desperately needed industrial production will be prevented from growing to its full potential because factories and businesses are forced or induced to locate in the wrong place' (Mills and Becker 1986:138).

One of the most visible strategies of the Indian government to tackle the rising problems of rapid urban growth in the large cities was the creation of new towns. Industrial towns (e.g. Durgapur) as well as administrative towns (e.g. Chandigarh, built after Independence) were built-up from scratch, some in the vicinity of existing large towns (e.g. New Delhi, New Bombay), others in the middle of rural area. Some of these new towns have succeeded in attracting a significant number of people and economic activities, others, especially those that were created in rural areas, have not been successful at all. At some point such new towns, especially those that are not located close to a large city, seem to reach a glass ceiling in their economic development, behind which they cannot grow any further. As a result, the overall effect of such cities on the traditional urbanisation pattern, heavily focused on the largest metropolitan areas, must be considered fairly limited.

Urban planning

Beside these moderate attempts to modify the urbanisation pattern, several urban policies were developed at the state and local level to tackle such metropolitan problems like housing shortages, problems of water and electricity supply, absence or inadequacy of a sewerage system, transportation bottlenecks, and other typical problems associated with rapid urban growth. Sometimes these policies were based on the general recommendations in the Five Year Plans, but more often they were simply *ad hoc* policies that differed significantly from one state and/or city to another.

So far as the Five Year Plans are concerned, nothing specifically was mentioned on urban development in the First Five Year Plan of 1951-1956. In the Second Plan (1956-61), the need for a planned development of cities and towns was recognised and the Delhi Master Plan was prepared as a kind of experiment in Indian town planning and redevelopment. Implementation of the plan was left to a newly established development authority: the Delhi Development Authority (DDA). The Third Plan (1961-1966), apart from recognising for the first time the role of industrialisation in the urban development process, also provided financial aid for the preparation of nearly 400 master plans and for the creation of a number of new state capitals such as Bhubaneshwar, Gandhinagar and Bhopal; new capital cities which had to be built following the reorganisation of states in 1956. Further, certain guidelines were put forward to solve the social problems of living in slums, which was a marked difference from the earlier policies of forced slum clearance. The Fourth Plan (1969-1974) recognised the need for financing urban development schemes, and the Housing and Urban Development Corporation (HUDCO) was established to provide the funds for metropolitan authorities to finance housing schemes. Further, state governments were advised to create planning and development authorities for the larger cities. The urban development chapter of the Fifth Plan (1974-1978) laid down a number of vague urban planning objectives (such as for example an increase in urban civic services), but was in general mainly outlining a limited and contradictory urbanisation policy.[14] The Sixth Plan (1980-1985) for the first time officially acknowledged the significant regional differences in urbanisation rates between states, and focused mainly, as previously in the fifth plan, on an urbanisation strategy for the country at large. The Seventh Plan (1985-

1990) gave somewhat more priority to rural development, explicitly stressed the necessity to slow down urban growth in the large cities, and promoted growth in the smaller cities and towns. It also advised state governments to provide more urban development funds to the municipal authorities. This plan also suggested land acquisitions around the large cities and the subsequent development of this land for urban use and resale, to generate the necessary finances to tackle the numerous urban problems and to prevent a rapid rise in urban land prices.[15] Finally, the Eighth Plan (1992-1997) has again been paying considerably more attention to urban planning and urban growth strategies, but since the change of government at the Centre in 1996, most attention has again been diverted towards rural development.[16]

A lot of confusion in urban planning has been caused by the unclear and unbalanced relations between the Centre and the states on the one hand, and between the states and the municipalities on the other. In the Indian Constitution, the exclusive power to enact legislation concerning urban policy and land policy was given to the states. In reality, however, all policy initiatives in urban affairs have come from the Union government, which curiously enough has a separate Urban Affairs Department. This department (through the Town and Country Planning Organisation) plays an advisory role in urban matters towards the states. Beside this department there is also the Planning Commission, another Union government institution, which plays a more direct role in the field of urban development, mainly through its national Five Year Plans.[17] Some of the guidelines in the Five Year Plans concern specific policies for urban planning.

On the other hand, the steady erosion of the functions and powers of local government in India and the gradual shift of political power from the municipal level to the state level since the mid-1960s, has been a major problem for most cities in India. Indeed, in most cities where large urban investment programmes have been implemented, urban development authorities (UDAs) have been set up, formed on the model of the DDA in New Delhi. These UDAs, as well as other state level organisations that were established over the years and are active in specific urban fields (e.g. HUDCO for housing), operate directly under the state government, and not under the local authority. In practice, the present situation in most large cities in India is such that the state authorities finance, co-ordinate and implement urban infrastructure works, whereas the urban tasks of the municipal authorities are strictly limited to maintenance activities. In many instances, however, the local authorities no longer have the financial means even for such limited activities. What metropolitan India needs first and foremost, are therefore clear demarcation lines of responsibility between Centre and states, and between states and municipalities, as well as strict co-ordination in financing, planning and management between the different political and administrative levels.

Conclusion

Most of India's Five Year Plans have started from an anti-urban bias, as if the growth of cities is the major evil which has to be extirpated. As we have said before, this conception diametrically opposes the natural tendency towards urbanisation and urban growth that any country at some point will experience in its modernisation process. Moreover, the urban management strategies that were put forward as guidelines to the states have seldomly been implemented, mainly due to lack of finances, confusion about administrative responsibilities, and/or political neglect. Looking ahead, there can be no doubt that the large cities of India will continue to grow rapidly, and that the urban problems will become worse and worse if full political attention is not radically shifted towards a proper national urbanisation strategy, as has been put forward separately by Renaud and Richardson, and towards urban management strategies for the large cities that can work in practice. If the national and state authorities continue to believe that India's urban problems can be tackled by promoting growth of small towns and villages, their naïvity could lead to a dramatic urban collapse in a number of cities and could obstruct the much needed large-scale modernisation of the country; a process from which the millions of urban poor will suffer the most.

As Mohan (1996) has correctly argued, India urgently has to search for methods that can cope with the growth of cities rather than to search for methods to stifle this growth. Given the enormous economic importance of the largest cities for the development of the country in general, he suggested that these cities should be regarded as national cities, and should be handled and managed as such. Internally, a policy of 'area specialisation' could be planned: an urban land use policy in which different city sectors *specialise* in one major economic activity, shifting most other economic activities away to other city areas in order to create an equilibrium in economic activity and land use in India's cities. So far as

Bombay is concerned for instance, it should not be impossible, given the right policies and necessary political determination, to shift everything but the financial sector out of South Bombay (all political activity to New Bombay, all manufacturing to the northern suburbs or beyond the city boundaries, etc.). In this age of modern telecommunication and information technology there indeed is no longer any meaningful argument for the government to be located in the same area as the financial sector. With this, however, we are already discussing urban development in the city of Bombay, to which we shall now turn.

Notes

[1] The first planned cities of China date from about a thousand years later.

[2] Although these have in a few words been the general theories about the Harappan civilisation for several decades, very recent findings have provided a lot of exciting new information on the age, the size and the nature of the civilisation, which may soon require a thorough revision of India's ancient history and urbanisation. For more on this see for instance India Today, 26 Jan 1998:pp. 44-51.

[3] Just before Independence, Bombay and Calcutta handled over two-thirds of India's trade, which is clear enough evidence of the enormous impact of the colonial port cities in British India.

[4] The negative growth between 1911 and 1921 was due to the world-wide influenza epidemic that followed the first World War.

[5] Today, only one country in the world, China, has a larger total population (1.234 billion) than that of India, but by 2025, when China is projected to have a total population of 1.526 billion, the gap between both countries will almost have been closed. China is nearly three times the size of India.

[6] Urban growth steadily accelarated between 1911 and 1951, slowed-down between 1951 and 1961, accelarated again between 1961 and 1981 and sloweddown again between 1981 and 1991.

[7] It must be noted that there is a marked difference between North India and South India in this respect. In the south, the proportion of entire families that migrate is, for reasons which are still not fully understood, much larger than in the north.

[8] Many of the generalisations on urban migration in developing countries, such as for example that all urban migrants are poor, do not hold for India (as they probably do not hold elsewhere). On the contrary, the proportion of rich people that migrate to Indian cities is relatively large compared to the proportion of poor people that migrate.

[9] Important in this, and indeed in the entire urbanisation process, is of course the way in which *'urban'* is defined. Since 1961, Indian censuses have been fairly consistent in their definition of an urban place. According to the Census, a settlement is defined as urban when its population is over 5,000, its population density is over 400 per hectare (or 1000/mile^2), and 75% of its male labour-force is engaged in non-agricultural activities. So far this definition is quite clear. In addition, however, as Mohan explains, a settlement can also be defined as urban by government notification, and the census authorities also have discretion to classify as *urban* 'some places having distinct urban characteristics even if such places do not strictly satisfy all the criteria mentioned above' (Mohan 1996:95). However, as Mills and Becker conclude, the Indian definition of *urban* appears overall to be 'as precise and satisfactory as that employed in almost any other country' (Mills and Becker 1986:33-35).

[10] The smallest states have been excluded for their size; Jammu and Kashmir has been excluded because the figures are estimates.

[11] This does not mean, however, that modernisation in agriculture will automatically lead to a replacement and decrease of agricultural labour, as the Indian example in recent times proves.

[12] While interviewing the Secretary Urban Development of the Government of Maharashtra (D.T. Joseph), I was told (to my surprise) that an overview of official State and Union Government legislation in urbanisation policy simply does not exist. So far as central policy is concerned, only recently a draft National Urban Policy has been prepared by the Town and Country Planning Organisation New Delhi (in 1992), and in Maharashtra one needs to go through the State's urban policy resolutions individually.

[13] In contrast to several developing countries, the physical closure of large cities to keep new migrants out has never been part of an 'urbanisation strategy' in India. Article 19 (1) of the Indian Constitution guarantees to every citizen 'the right to move freely throughout the territory of India' and 'to reside and settle in any part of the territory of India'. A total ban on migration of people, however objectionable from a social point of view, can thus neither be sustained constitutionally.

[14] For instance, on the one hand the development of small towns and new urban centres was promoted, whereas on the other hand inter-state metropolitan projects were to be developed.

[15] The policy makers must have realised that what they were suggesting was quite contradictory in nature.

[16] For a detailed account of urban planning guidelines in the Five Year Plans see Ghaneshwar (1995).

[17] The 'federalisation' of state subjects is a typical Indian phenomenon and is largely a result of the centralisation of political power in Delhi since the rule of Ms. Indira Gandhi.

4 Urban Development in Bombay

Introduction

In the previous chapters, we have been looking at the urban growth and urbanisation processes in the Third World in general, and in India more specifically. It is now time to turn our focus to developments in the city of Bombay. In this chapter we will first look at the historical growth process of the city, during the colonial period and thereafter since Independence. In doing so, attention will mainly go to the different economic and political factors that were important in this process and that have caused dramatic changes in the demographic and spatial urban patterns. Secondly, we will look at the consequences of Bombay's very rapid growth, and we will discuss the debate of the 1950s and 1960s on which policies to be followed to prevent the city from the 'unavoidable collapse', which was generally expected at the time.

First, however, we shall briefly introduce the nomenclature of geographical and administrative delimitations that will be used throughout the text.

1. *Bombay Island*, also called *South Bombay* or *Bombay City*, is the original island where development of Bombay started. Today, after the implementation of a series of reclamation schemes, it covers an area of 78 km² of land, concentrating the major part of all modern urban activities such as business offices, administrative offices, industry, culture, etc. The area is also very intensively used as a residential area.

2. The *Bombay Suburbs* is an area of 525 km² that rapidly started to develop after Independence, mainly due to the gradual congestion of Bombay Island. Whereas during the colonial period this area was mainly used as agricultural hinterland for the island's population, at present it is the most populated area of Bombay.

3. Bombay Island and Bombay Suburbs together are called *Greater Bombay*, which thus covers an area of 603 km², and with that most of Salsette Island. Since 1950 the Greater Bombay Municipal Corporation (GBMC) has been the political authority for this area, which in terms of land revenue and general administration is regarded as one state District (out of the 27 in which Maharashtra is divided). Greater Bombay is further

subdivided into wards and districts: 7 wards (38 districts) on Bombay Island and 8 wards (50 districts) in the Bombay Suburbs.

Figure 4.1: The Bombay region (1=South Bombay or Bombay Island/2=Bombay
Suburbs/1+2=Greater Bombay/3=BMR/4=New Bombay)

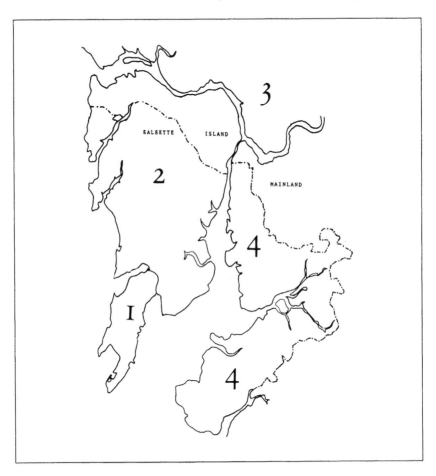

4. The *Greater Bombay Urban Agglomeration* (GBUA) is an area of 1178 km², for the first time included as such in the census of 1991. It covers the area of the GBMC + the total area of five adjacent municipalities on the mainland.
5. The Regional Plan of 1970 has created the *Bombay Metropolitan Region* (BMR), basically as a basis for regional planning. It is a large area of about

4375 km² which includes parts of three different state districts, i.e. Greater Bombay, the remaining land of the Salsette island north of the suburbs, and large adjacent areas on the mainland which are being administered by several municipal corporations and a number of municipal councils and villages. A relatively small part of this area is the twin-city of *New Bombay*.

Growth of Bombay in historical perspective

Bombay, mid-17th century: seven small islands off the West Coast of the Indian subcontinent, surrounded by marshy land; few pieces of fertile land covered with palm-trees; clusters of villages inhabited by *kolis* (fishermen) and farmers cultivating some rice, vegetables and fruits for personal consumption; no commerce; no important rivers; a rather infertile Konkan hinterland with no mineral resources, no artisan industry, and hence no meaningful economic activity.

Bombay, late 20th century: one of the biggest metropolitan areas in the world; about 15 million people; India's commercial and financial centre, having the most important port, the busiest international airport, the highest concentration of industry and the highest concentration of multinationals in the South Asian region; political capital of the state of Maharashtra, a state bigger than the UK in terms of area as well as population; and one of the major centres of culture on the subcontinent.

Bombay, 350 years of dramatic growth and expansion. How could a region with apparently so little potential and within such a relatively short time span become India's number one city and one of the world's most populated urban areas? This is the question we shall be looking at in the next section.

Growth of Bombay during the colonial period

Early growth of Bombay   When the Portuguese in 1661 gave the seven islands as part of a dowry to England's Charles II, the total number of people living on the islands was about 20,000 (Kosambi 1986:35).[1] Four years later the British government took possession of the islands, considered them rather useless, and leased them to the East India Company (EIC) for an annual £10 Sterling in gold.

The EIC, already having a major trading post in Surat on the Indian West Coast, initially used the islands during the second half of the 17th century as a mere geographical extension of its commercial activities in Surat.[2] Bombay had no real economic resources and did not seem to have anything worthwhile. Its location, however, had two important potential advantages: safety against possible enemies, and a natural harbour with exceptionally deep water. Because of these advantages the EIC ultimately decided to shift her registered office from Surat to Bombay. Until the mid-18th century, Bombay nevertheless remained very dependent on the trading activities in Surat. But with the gradual decline of the Mughal empire and the growing instability in a number of Princely States in the north, many Gujarati traders and artisans left Surat and moved down to Bombay. Among them the Parsi community, who had accumulated a significant trading capital. Their commercial experience and financial strength were to become a crucial catalyst for Bombay's initial growth and development.

From the mid-18th century the importance of Bombay as a major seaport was growing fast and the population rapidly increased to 100,000 by 1780 (Harris 1978:table 3). Another important element in Bombay's initial development was the beginning of the first Industrial Revolution in England. Bombay happened to be the nearest seaport to the cotton growing areas of what is now roughly the state of Gujarat, and hence became the major channel through which raw cotton was shipped to Europe.

From the very beginning, Bombay's urban layout was characterised by a clear ethnic and racial segregation, with the Europeans living in spacious bungalows in a small, but well-planned fortified area in what is known as the *Fort* area, where all the colonial activities were centralised. In the periphery, Bombay started to grow in an uncontrolled way. The indigenous Maratha people were pushed northwards by the more well-off Indians, and houses were gradually being constructed further and further away from the Fort area. The marshy land in-between the five northern islands was drained and hence became one vast landmass by the end of the 18th century. The two small southern islands were transformed into cantonments.

From the earliest days, Bombay Island had a kind of symbiotic relationship with Salsette Island in the north, the largest part of which later would become known as 'Bombay Suburbs' and would become incorporated within the administrative and jurisdictional boundaries of Greater Bombay.[3] Until the late 19th century, however, this Salsette Island largely remained a rural area, was mainly used as agricultural hinterland, and functioned as an excellent buffer zone against potential invaders.

Bombay during the 19th century   The 19th century is of crucial importance for a good understanding of contemporary Bombay. It was in this period that Bombay developed from a stagnating culture with slow pulse into an economically thriving and, for its time, modern city.

All through the 19th century, internal as well as external political and economic events had a major influence on Bombay's growth and development. In 1813 the Charter Act abolished the monopoly position of the EIC and Bombay's market was opened up for private traders and missionaries. Some years later, at Kirkee, EIC troops defeated the Maratha powers on the mainland and annexed most of the Maratha dominions. Consequently, the EIC from now controlled the major part of western India, which meant that a large new market was waiting to be exploited and that the cotton fields came under direct control of the EIC. These newly incorporated Maratha regions were grouped into the 'Bombay Presidency', with Bombay City as the new capital of administration.[4] Hence, from its isolated geographical position on the Indian West Coast, Bombay became the political centre of western India.

Trade and commerce remained Bombay's core activities. The city's import and export patterns were typically colonial, with raw cotton and opium being shipped out of the country, and cheap industrially produced British textiles coming in. From the mid-19th century, however, Bombay itself entered the industrial era. Making use of imported British technology and supported by the availability of an indigenous capital, Indians themselves started to invest in a cotton-based industry 'to fight Manchester with her own weapons' (Edwardes 1902:264). The first cotton factory in Bombay was set up in 1854, and thirty years later there were 35 factories employing over 30,000 labourers (Harris 1978:8). Due to the flourishing cotton industry, new capital was attracted from all over the Indian subcontinent, shares on the Bombay stockmarket reached new heights, and other industries were being established. Logically, the construction of an extensive transportation network, inland as well as overseas, was of crucial importance to Bombay's business class. From the moment the Maratha dominions were annexed, these business circles pressurised the EIC to provide proper communication links. Soon, an extensive road network connecting Bombay with the mainland was being established, and the first railway link on the Indian subcontinent was built between Bombay and Thana, 32 kms further north. The cotton fields now came within easy reach of the factories. Furthermore, when in 1869 the Suez Canal was opened,

Bombay also became the nearest Indian port to Europe; a key advantage over the other major ports of Madras and Calcutta.

Because of these developments, Bombay's industrial growth potential was now enormous. Although the region had a serious disadvantage in that it lacked coal mines and basic metals, all the other necessary elements to make it a thriving industrial area were available: an experienced (mainly Gujarati) entrepreneur class, a large amount of indigenous capital, a large supply of cheap labour, excellent transport facilities, and an almost unlimited supply of raw cotton. Consequently, Bombay's existence became much less dependent on overseas trade than hitherto, which seriously increased the economic stability of the city. The city's role as an exporter gradually shifted into that of an importer for the development of a small, modern economy; a role which according to Harris (1978:7) clearly pointed towards an increasing integration of Bombay in the Indian economy.

In the meantime, the British government increasingly exercised control over the entire subcontinent. In 1958, after the Mutiny the previous year, the British Crown took over control from the EIC.[5] Consequently, rule over Bombay also shifted to the British government, and the role of the city for the Empire would become increasingly important. During the 1870s, a number of Bills resulted in the establishment of the first elected city council, the Bombay Municipal Corporation (1873), which became responsible for typical municipal services such as water supply, drainage, education, health care, and the construction of streets and bridges.[6]

Demographically, natural hazards and certain political events were sometimes having a serious impact on the population pattern. In 1802 for instance a drought destroyed the crop and caused severe famine in the countryside, after which thousands of people fled to the major cities in search for food. And when in 1818 the EIC troops defeated the Maratha powers, masses of Maratha people moved from these areas to Bombay. Nevertheless, in terms of population growth such events were of minor importance compared to economic motives. Any spurt in Bombay's economic activity always went hand in hand with in-migration, especially after the technological revolution in transportation, which removed the practical impossibilities for people to do so. Hence, parallel with Bombay's economic growth, the population increased fairly quickly from an estimated 221,000 in 1815 to 822,000 by 1891 (Harris 1978:97).

Table 4.1: Population increase Bombay City, 19th century

1780	100,000	1846	566,119
1790	113,726	1864	816,562
1814	170,000	1872	644,405
1830	229,000	1881	773,196
1836	236,000	1891	821,764

Source: Harris (1978)

The main characteristic of 19th century Bombay's population was its diversification, which crossed racial, religious and linguistic lines. Many demographic characteristics were quite typical for the three main Anglo-Indian port cities of Bombay, Calcutta and Madras. First, the European elite was very small in number and to a large extent dependent on the co-operation of different layers of Indian society. Secondly, every community had its place in the occupational structure. Caste and language were therefore already a good indication of someone's job. And thirdly, Bombay's population was, in accordance to Horvath's model of colonial social stratification (Horvath 1969:69-82), roughly structured around three main strata: the European elite, an intermediate group of Gujarati traders and Parsi entrepreneurs, and the local Maratha community.[7]

In the course of the 19th century a number of very important new elements were added to the city's spatial pattern. A first new element was the creation of a second Native Town, about a mile north from the old one, which would soon become a second urban and economic core. Other elements that were added to the urban landscape were the connection of the two small southern islands to each other and to the rest of the island, the piers and docks that were built on the eastern side of the island, and the new Cotton Green that was created on the western side.

In the north and the west, the first so called 'garden suburbs', spacious European style suburban areas of a pre-industrial nature, were being laid-out.[8] An increasing number of Europeans escaped from the heat, the dust and the indigenous population in the Fort area to these hilly green suburban regions, where they could live in spacious houses, build their own clubs, and relax in their large gardens. At present, these areas are still the residential zones of Bombay's upper class. The former European residential areas in the Fort were gradually commercialised and replaced by a Central Business District (CBD), where all the important commercial and administrative functions were centralised. Other new and important spatial developments were directly connected to the process of industrialisation. A first series of cotton mills were built close to the new Native Town, a second in the northern part of the island which was now by rail connected to the southern part. This was the beginning of a rapid industrial development in the northern districts of the island that was soon to become an important second economic core. Overall, one of the most distinct features of Bombay's early spatial pattern throughout the 19th century was that urban growth and development were almost strictly limited to the southern half of the island. The northern half remained largely rural and sparsely populated by contemporary standards.

Because the need for more space in the southern part was high, the walls of the Fort were demolished and land reclamation schemes were being implemented. The newly available space was mainly used to erect magnificent governmental and other buildings such as the High Court, Bombay University, the General Post Office or Victoria Railway Station. These new buildings and the European garden suburbs were the clearest expression of the city's growing wealth and prestige. Nevertheless, the large majority of the population lived in miserable conditions, in the crowded Native Towns and, since the industrialisation, in so called *chawls*, in which the industrial workers, close to their factories, were pushed together. Apart from this, a growing number of people were 'floating' and the number of slums was constantly increasing (Ramachandran 1977:42). Basic facilities such as running water or a sewerage system were almost absent and not available for ordinary people (Harris 1978:10). Bombay's pride was, as Harris has put it, 'in achieving the first elected Municipal Corporation, India's first stock exchange, private telephones and tramways... rather than in the use of its wealth to improve the welfare of the majority' (Ibid: 11).

Bombay from the late 19th century until Independence By the late 19th century, Bombay's dominant position in the West Indian region had become unchallenged, and the role of the city in the Empire of great importance. It was the functionally most diversified city, with a concentration of commercial, industrial, administrative and cultural functions, and had a population six times the size of the second largest city within the Bombay Presidency. In the period from then until Independence, this process of rapid urban growth and development continued. We will now briefly consider the political and economic developments and the evolution of the demographic and spatial patterns of the city during this 50-year pre-Independence period.

After the turn of the century, Bombay's most important function in economic terms remained overseas trade, which since 1873 had been completely controlled and managed by the Bombay Port Trust (BPT). During the first decades of the 20th century, the trade pattern remained typically colonial, with raw cotton still being the most important export article.[9] Nevertheless, by the 1920s, Bombay's import and export pattern would have become much more diversified. It was no longer mainly focused on the UK as it had been in the 19th century, but also to an increasing extent on Japan, China, Continental Europe and the US (Sharpe 1930:appendix-E).

Although trade remained Bombay's most important economic activity, growth in the industrial sector was rapid. The indigenous investment pattern, which was a very distinct feature of Bombay's economy compared to most other major Indian cities, was being continued. Consequently, by the turn of the century Bombay contained 82 cotton mills employing 73,000 workers. By 1908 the number had gone up to 166 factories (predominantly cotton mills), and by 1921 the number of textile workers had doubled to 146,000 compared to 1900. This was over 11% of the then total population (Harris 1978:8). Around 1910 a first cotton mill was set up on the southern tip of Salsette Island, just across the northern border of Bombay Island. It was the first step in a process of wholesale industrial and demographic expansion on Salsette Island. In view of this gradual but steady suburban growth, the Bombay Suburban District was established in 1920, which after Independence would be replaced by the GBMC.[10] Other new elements to the city's economy were the first multinationals that were entering Bombay Island from the early 20th century. Overall, it was not only the import/export pattern which became more diversified during this period, but in fact the entire city economy. As a consequence, urban stability again increased. But, as events in 1929 proved, the city's economic base would still not be invulnerable. Due to the world-wide economic-financial crash, the number of male industrial workers in Bombay decreased with nearly 20% and the number of female industrial workers with nearly 50%.

In this period as well, the transportation sector would play a crucial role in Bombay's urban growth.[11] Since the railway links between Bombay and the mainland were crossing Salsette Island, people were soon settling around the suburban railway stations. Hence, small clusters were formed, which after Independence would develop into real urban centres. Both, the establishment of cotton mills in combination with the railway lines between

Bombay Island and Salsette, were thus the main catalysts of the urban sprawl into Salsette Island and the growth of the 'Bombay Suburbs'. Some time later (1926) the first English double-deckers were introduced in the urban landscape, and by the time of Independence 150 buses (carrying 100 million passengers per year) were servicing the city.

Apart from two serious but temporary set-backs, a plague epidemic at the turn of the century and the financial crash and recession of the late 1920s/early 1930s, Bombay's demographic pattern showed a steady, though fairly slow population increase throughout this period. From 822,000 in 1891, the population went down to 776,000 by 1901 (plague), but quickly recovered to nearly 1 million by 1906, and went further up to 1.18 million by 1921 and to nearly 1.5 million by 1941 (Harris 1978:table1.3).[12] Nevertheless, over a period of fifty years, Bombay's population had not even doubled.

Table 4.2: Population increase Bombay, first half 20th century

1891	821,764	*1921*	1,175,914
1901	776,606	*1931*	1,161,383
1906	977,822 *	*1941*	1,489,883
1911	979,445		

* Special census. Source: Harris (1978)

As discussed in previous chapters, total urban growth in a city is the sum of natural increase, net migration, and the impact of changing city boundaries. It has been argued by some that the rapid urban growth process in the Third World in general has been more due to natural increase than to urban migration. For the period discussed here, this is certainly not correct as far as Bombay is concerned. Between 1881 and 1931, only one in four *Bombayites* had been born in the city, and in times of exceptionally high economic progress and prosperity (as in 1921) the ratio went down to even one in six (Crook 1993:121, Harris 1978:28); a pattern clearly expressed by the very unequal sex ratio in the city in that time. In 1881 there were 151 males per 100 females, and by 1921 this had gone up to 191 males per 100 females.[13] The majority of migrants came from the most backward areas within the Bombay Presidency itself, which points to a strong push factor in the migration process (Harris 1978, Yadava 1989). The largest share of migrants coming into the city in those days were unskilled labourers, which indicates the importance of the booming industry, and points to a strong pull factor in the migration process. Thus, for Bombay it was certainly true that 'the faster the pace of industrial growth, the more rapidly the native-born share as a proportion of the total population in the city declined' (Harris 1978:10).

The relationship that colonial Bombay, throughout its history, has had with its periphery has been of crucial importance in the migration process. Unlike the other major colonial port cities, Bombay hardly contributed to the development of its hinterland. On the contrary, it gradually became an economically important city on the back of its periphery, focusing overseas instead of inland. Since Bombay became increasingly wealthy and its rural hinterland increasingly poor, it is apprehensible that an increasing number of poor agricultural labourers and unemployed fled to the city to find a job in industry. This process obviously caused serious disturbances in the subtle socio-economic balance in the countryside as well, often resulting in even more poverty and more urban migration.

As might have been expected, the large influx of migrants from the countryside in Maharashtra as well as from other states in India (primarily Tamil Nadu, Kerala and Sind), did not have a major impact on Bombay's pattern of social stratification. Migrants coming into the city were being absorbed by their own religious and/or caste community. Hence, Horvath's model of social stratification remained valid for Bombay's population up and till Independence, with the Europeans monopolising the top jobs, the Gujaratis and Parsis having commercial functions, and the local Marathas working as clerks or in industry. Overall, the entire population remained highly diversified but strictly classified according to job, social position and spatial ordening.

Bombay's spatial pattern during the early 20th century was a continuation of the basic, Indo-British pattern that had already been laid out by the late 19th century, with the Fort area slowly being transformed into a Central Business District, a large Native Town, spacious European suburbs outside of the crowded areas, a cantonment in the south, and two industrial zones surrounded by agricultural land in the largely rural northern half of the island. In the period up to Independence, the European suburbs on Bombay Island would gradually become less exclusive and local elites were now also moving into these areas. The number of reclamation schemes increased, especially on the eastern side of the island, where by the late 1920s not less than 1880 acres of land were won over the sea, in that time about 1/8th of the entire island (Sharpe 1930:47-57).

The overall pattern of population density remained a contrasting one between the northern part of Bombay Island and the southern part, between a largely rural half and an urban half. The population was indeed spreading northwards, but population density in the northern half of the island remained, throughout this period, relatively low (figure 16). In terms of housing, the main feature was that the number of *chawls* and slums constantly increased.[14] The early phases in the process of urban sprawl onto Salsette Island were, as we saw, another new element in Bombay's spatial pattern during this period; an urban sprawl which was slow but would gradually bring new socio-cultural and economic aspects to urban life.

Growth of Bombay after Independence

The first 25 years after Independence was the period in which the idea to develop the vast area across the island as a 'counter-magnet' to Greater Bombay gradually assumed concrete shape.

Political evolution   When India won its Independence in 1947, it was at the same time forcibly divided into two new independent states: Pakistan (West and East) and India. Millions of Hindus fled the newly established Islamic state of Pakistan, primarily to one of the large Indian metropoles. Hundred of thousands of refugees arrived in Bombay coming mainly from Sind, formerly part of the Bombay Presidency. They were accommodated in a refugee camp in Ulhasnagar, some 35 miles northeast of Bombay, which later would develop into one of Bombay's major satellite towns. The former Bombay Presidency was obviously dissolved, and the new 'Bombay State' was created. In 1961, in accordance with the federal government's policy of reorganisation of the Indian states, this Bombay State was divided into the two new states of Gujarat and Maharashtra. Bombay however retained its former function as the political capital of the new state of Maharashtra; a function, which would become increasingly important, not least in terms of employment.

Economic evolution   In economic terms, the general tendency after Independence was an increasingly speedy growth and diversification of industry, trade and services, and because of a gradual economic and urban congestion, a rapid geographical expansion of activities to areas beyond Bombay Island. During the first post-colonial years, this process was limited to the suburbs of the Greater Bombay area, but later also crossed the administrative limits of Greater Bombay and spilled over to the northern parts of Salsette Island and beyond, onto the mainland north and east of Bombay.

The economic transformation from a traditional textile industry into a modern urban industry had already started in the early 1930s, but was fairly slow compared to the growth in the tertiary sector. From Independence, the industrial sector modernised and diversified much more rapidly, and became an increasingly important contributor again to the city's economy. Between 1951 and 1971, 42% of all newly created jobs in Bombay were to be found in industry (Harris 1978:13), and by 1971 no less than 20.7% (over a million workers) of India's total factory employment was located in Maharashtra state (Statistical Outline 1995:149), the majority of which in Bombay and adjacent areas (Harris 1978:20). Employment in industry was in Bombay in fact higher than employment in trade and services.[15] Moreover, by the early 1960s, industry contributed an exceptional 56% to the city's income (Harris 1978:16), and in terms of industrial production, 24% of India's total output in the mid-sixties was produced in Maharashtra (81% of which was produced within the Bombay region) (Harris 1978:18,20). In the decades thereafter, Bombay would remain the prime industrial city of India. By the early 1990s, 15% of all factory employment in India was still concentrated in the state of Maharashtra, which produced nearly 22% of total national industrial output (Statistical Outline 1995:87,149). For both, state industrial employment and output, a large part was still generated in or around Greater Bombay (*Selected Indicators for Districts in Maharashtra and States in India 1990-91*:33).[16] Within the industrial sector, the traditionally strong textile industry has been in a relative decline compared to other industrial sectors, but it remained the largest job provider in industry till the 1970s.

Light engineering and petrochemicals have become increasingly important, both in terms of production and employment. By the early 1970s, Bombay accommodated an enormous 60% of the total number of small-scale industrial units in the state of Maharashtra, which had the highest number of these units in the country (Harris 1978:15). The increase in employment in the capital-intensive chemical and petrochemical sectors was obviously much smaller than in other industrial sectors, but the increase in output and its share in the city's overall produce has, nevertheless, been very significant and increasingly important.

Finally, Bombay's commercial film industry is occupying a distinct position in the city's industrial sector. Bombay has one of the largest film outputs in the world, and the film sector contributes significantly to the city's capital and employment. However, an exact figure of the contribution of the city's movie industry to the overall urban economy is not available, and is due to the large amounts of black money in the movie business difficult to assess.

Nevertheless, the figures above are clear evidence of the important role of industry in Bombay, in earlier as well as in more recent times. Other sectors however have been of almost equal importance. Output and employment in the traditionally strong trade and services sector has significantly increased since Independence, and was no longer almost exclusively focused on the export business. By the early sixties, 20 to 25% of Bombay's income was generated by its domestic trade, banking and commerce (as compared to the share of industry in this income of 56%). The services sector thus became one of the most important components in the city's prosperity.[17]

The financial component has been occupying an increasingly important position. The Stock Exchange and the Reserve Bank of India had both been established in Bombay in the 1930s, and in the post-colonial period they attracted numerous other banks and insurance and investment companies. In turn, the latter have acted as catalysts for wholesale and retail trade, social services, construction, urban transport, and so on. Consequently, Bombay also became the financial nerve centre of the country. Already by 1972, 70% of India's bank deposits were held by banks with headquarters in Bombay (Harris 1978:17). This tendency has continued ever since. By 1993, the number of cheques cleared in Bombay was more than double the number cleared in Delhi, which by then had become the second financial city in India, and the amount of cheques cleared was more than seven times as high in Bombay as in the capital (Statistical Outline 1995:210). In terms of revenues, Bombay, by the late 1980s, accounted for about 40% of the income tax and 80% of the state sales tax collections (Sundaram 1989:21).

As a result of this rapid economic growth and diversification, Bombay developed into the busiest seaport and the major airport of the country; a trend which would continue throughout the 1970s, 1980s and 1990s. By the early 1970s, 38% of domestic and 57% of international air traffic in India passed through Bombay (Harris 1978:16), and by the early 1990s, passenger air traffic through Bombay was 59% higher and cargo air traffic 28% higher than through Delhi Airport, which by then had become India's second most important airport (Statistical Outline 1995:209).

Finally, there is Bombay's informal sector, which has experienced a massive growth since Independence. As the organised sector mostly requires an education or specific skills, it is the informal sector in which most migrants hope to find employment (Gupta 1985:179).[18] How large exactly the informal sector in Bombay is, is obviously difficult to say, but most estimates only start from 50% of total urban employment. All in all, by 1981 Greater Bombay's official employment figure was 2.9 million (Census 1981), about 24% of which was in the public sector (Sundaram:23), and by 1991 the official figure would have gone up to 3.5 million (Census 1991).

Demographic change   Whereas the long-term increase in population in colonial Bombay had in fact been fairly slow, from 1947 onwards economic diversification, growth and employment were accompanied by a tremendous increase in population, as well as by unprecedented physical expansion.

For so far as Bombay City is concerned, the population increase on the island of 56% (1.49 million to 2.33 million) during the decade 1941-1951 was the sharpest increase ever. In the decades thereafter growth on Bombay Island would slow down, mainly due to saturation, and would ultimately become slightly negative during the eighties.[19] The decade 1941-51 however was also the period in which the population in the suburbs started to explode, and the small centres around the suburban railway stations started to develop into real towns with a typical urban character (Rajagopalan 1962:67). Between 1941 and 1971 the suburban population doubled in each decade, from 0.20 million in 1941, over 0.62 million and 1.38 million in 1951 and 1961, to 2.90 million by 1971. Population growth in the suburbs has also continued over the years. It has slowed down here as well, but the increase in absolute figures has been enormous, from a mere 200,000 in 1941 to over 6.7 million by 1991.

In 1991 the Greater Bombay Urban Agglomeration (GBUA) was formed, which consists of the GBMC (9.92 million) and five additional municipal corporations on the mainland (2.67 million in total). This makes that Bombay's total population in 1991 was in fact not just under 10 million, but about 12.5 million, and that the city was occupying the sixth rank on the table of most populated cities in the world, after Tokyo (around

25 million), New York (16 million), Mexico City (15 million), São Paulo (14.8 million) and Shanghai (13.5 million) (UN 1996:32-33).

Since about five years, it is actually not Greater Bombay that is growing fastest, but the satellite towns, new towns and further-off suburbs outside the GBMC area, which mainly act as dormitory towns. Thus, whereas it may seem that population growth in Bombay during the 1980s, with a growth of just over 20% between 1981-91, is being brought under control, the truth is that the metropolis is still growing at an alarming rate, only now primarily in the suburbs beyond the Greater Bombay limits. Recent UN estimates expect the GBUA to have 24 million people within 15 years and to move further up from the sixth to the third rank on the table of most populated world cities (United Nations 1994).

Figure 4.2: Population increase Bombay City and Greater Bombay, 1911-1991

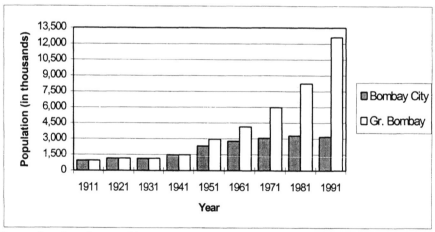

Source: based on figures table 4.3; Notes: Area extensions: 5 municipalities in the suburbs (176 km²) incorporated within Bombay area in 1950; 26 villages of Thana District incorporated in 1956 and total area for Greater Bombay becomes 438 km²; city boundaries extended for 3rd time in 1965 and total area becomes 603 km² (78 km² of Bombay Island + 525 km² of Bombay Suburbs). 1991 figure for Greater Bombay in graph above is figure for the GBUA (included in census for first time). The total area of the GBUA is 1177.55 km² and is the sum of the Greater Bombay area (603 km²) + the areas of Municipal Corporations adjacent to Bombay and belonging to the constituency of Thana District, i.e. Thana MC (144 km²), Kalyan MC (225 km²), Ulhasnagar MC (22 km²), Mira-Bhayandar MC (79 km²), and the part of New Bombay falling under Thana District (104 km²) (Census 1991).

**Table 4.3: Population figures and growth figures Bombay City,
Bombay Suburbs and Greater Bombay 1951-1991**

	City	Gr.	Suburbs	Gr.	Greater Bombay	Gr.	GBUA
1941	1,490,000		200,000				
1951	2,329,000	56%	627,000		2,956,000		
1961	2,772,000	19%	1,380,000	120%	4,152,000	40.5%	
1971	3,070,000	11%	2,902,000	110%	5,972,000	43.8%	
1981	3,285,000	7%	4,958,000	71%	8,243,000	38,0%	
1991	3,175,000	-3.5%	6,751,000	36%	9,926,000	20.4%	12.6 Mln

Source: Census of India 1951-1991 and Harris (1978). Figure for Suburbs 1941 from Correa *et al.* (1967:49). Note: population Greater Bombay = population City + population Suburbs. Gr. = growth.

Which factors have contributed to this massive population increase? In accordance with the urban development process in India in general, urban migration has clearly been the dominant factor in Bombay's urban growth process since Independence, but only until the 1980s, after which natural increase became the dominant factor. In 1941 not less than 72% of the city's inhabitants had been born outside the GBMC. In the decades thereafter that percentage went down to 64% (1961), to 54% (1971) and to exactly half (1981). Recent studies have demonstrated that natural increase presently contributes about 40%, migration about 35% and annexation about 25% to the city's population growth (Mukherji 1994). In absolute figures however the proportion of migrants in Bombay's population has constantly increased since Independence: from 2.13 million (1941-51), over 2.66 million (1951-61) and 3.22 million (1961-71), to 4.12 million (1971-81). Nevertheless, according to J.B. D'Souza, during the 1980s 'only' about 60 families migrated into the city every day, as compared to the official estimate of around 300 (D'Souza 1986-a).[20] Studies have proved that rural poverty and urban employment opportunities in Bombay's industry and in the informal sector have also been the determining factors in the migration process after Independence. The 'Bombay Slum Survey 1976-77' for instance found that only slightly over 4% of the city's total slum population had been born in Bombay, and of the 96% of migrants in the slums 78.5% gave 'potential employment' as the main reason for migration (Mayur and Vohra 1985:176).

From 1947, there were some clear differences in the demographic composition of the population in City and suburbs. Whereas the demographic pattern remained fairly stable in the City, in the suburbs it was striking that the presence of young males (20-40 years old) sharply increased, whereas it slowly decreased on the island. Primarily this was a result of the rapid industrial expansion in the suburbs (by 1960 the eastern suburbs already counted 238 industrial units and the western suburbs 198), and of the existence of good commuting facilities. The fact that by the late 1950s about 800,000 people commuted every morning from the suburbs to the City (Rajagopalan 1962:106) indicates that during this period these suburban areas remained largely satellite towns.

Evolution in the spatial pattern   Economic expansion and the rapid increase in population in the suburbs after Independence obviously resulted in an enormous dynamic with radical changes in the spatial pattern. On Bombay Island the former contrast between a predominantly rural northern half and an urbanised southern half would soon disappear, but the most dramatic developments occurred in the suburbs.

On Bombay Island, where more land was reclaimed, the original CBD in the Fort area became completely congested and a second major CBD was created around Nariman Point in the southern area. The increase in tertiary activities pushed up land prices, and industrial activities and workers were increasingly driven out to the northern districts of the island and into the suburbs. In residential terms, the former exclusive European areas were taken over by a westernised Indian elite, and the colonial residential segregation along religious and ethnic lines was soon after 1947 gradually transformed into a segregation along religious and socio-economic lines.

Changes in the spatial pattern after 1947 were much more radical in the suburbs, which since the early sixties have almost exclusively been responsible for Greater Bombay's overall population growth. Soon after Independence (in 1950) the GBMC was established, but population growth and dispersion in the suburbs were so rapid that soon afterwards (in 1956) the municipal limits had again to be extended northwards and 26 villages of Thana District were incorporated within the GBMC. Yet, population growth in the suburbs continued. Between 1961 and 1971, for instance, growth in some suburban districts separately was higher than the entire growth on Bombay Island over the same period (Harris 1978:21).[21] In contrast to what could have been expected, the fastest growth areas in the suburbs were not the southern suburban districts close to Bombay Island, but the north-eastern districts furthest away from the City. The main reasons for this were the vicinity and easy access to the expanding industries in the northern suburbs and in the Thana region north of Bombay on the mainland, and the lower quality of land and therefore lower property rates.

This dramatic redistribution of people and employment in Greater Bombay during the first decades after Independence followed roughly the classical development pattern of most industrial cities: a growing number of jobs and a refinement of functions in the services sector coinciding with a redistribution of labour-intensive activities to the suburbs. The diversification of industrial production and the congestion in the older industrial areas simply pushed the new industries to the edge of the city or beyond. Hence, from the early sixties three separate cores, with a very distinct density and land use pattern, were existing close to each other: the financial and services centre in the former Fort area in the southern tip of Bombay Island, the old industrial centre between city centre and the northern districts of the island, and the new industrial zone on and beyond the northern administrative borders of Greater Bombay.

In short, the rapid increase in population in Bombay was a clear reflection of the city's economic success. Unfortunately, that success prompted serious problems as well, to which we now turn.

Problems related to urban growth and policies for the future

Bombay, the state of Maharashtra, and in fact the entire country have benefited tremendously from the city's economic progress and wealth. Bombay's story, however, has not only been a story of progress and prosperity, but since Independence increasingly just as much a story of serious urban deficiencies and congestion.

Massive urban problems

As has been the case in most Third World cities, the consequences of Bombay's dramatic growth have been very significant. The urban crisis that first swept the city in the early fifties has worsened ever since. The most visible features of this crisis are the problems of shelter, transportation and insufficient basic services and infrastructure. At not a single point in Bombay's history have the city and state authorities been able to provide sufficient basic services to the growing population, and the urban managerial performance has been so poor that already since the late

1950s there has been a general fear that the city could collapse under the pressure. Apart from the city economy, the main victims of the urban crisis have been Bombay's inhabitants, predominantly the urban poor. They are the millions that have suffered most from the negligible amount of public housing, from the highly insufficient supply of potable water, from the absence of a properly working sewerage and sanitation system, from the poor education and health facilities, and from the absurdly overcrowded commuter transportation system. The figures hereafter will clearly picture the extent of Bombay's urban 'inadequacies'.

Housing and basic urban services   Throughout its history, Bombay has always failed to provide sufficient housing for its constantly increasing population. Already in 1872, the author of the Census observed that 'The houses of Bombay are really far too few in number to afford proper accommodation for its inhabitants...'. Since Independence, however, the situation has gone from bad to worse, has increasingly affected the middle classes as well, and has in general been far more acute than in most other Third World cities. Two particular Acts and a number of land use regulations have been of crucial importance in this deteriorating process: the Rent Control Act (RCA), the Urban Land Ceiling and Regulation Act (ULCAR Act), and the regulations concerning the use of Floor Space Index (FSI). They have created an overall situation of extreme shortage of private land and housing, resulting in astronomical land prices and property rates (see also chapter 9). As a result, Bombay is currently one of the most expensive cities in the world in terms of land value. What is of concern here, is the general impact of this housing shortage and the highly insufficient supply of other basic urban services.

So far as the formal housing market is concerned, the scarcity of land and the low supply of additional new housing (the latter which is mainly being constructed in the extended suburbs and satellite towns, far from the place of work), have resulted in a shift of population northwards and in hours of daily commuting on extremely overcrowded suburban trains. On the other hand, the large stock of dilapidated buildings on Bombay Island (a direct result of the RCA) has resulted in a qualitatively very poor living environment. A study by the Tata Institute of Social Sciences has revealed that no matter how dilapidated these structures are, no less than 60% of people occupying them belong to a minority of middle and upper middle income groups, thus pointing to the scarcity of housing. The '1967 Development Plan' already identified an estimated shortage of over 130,000 housing units, and the Bombay Metropolitan Region Development Authority (BMRDA) has put the housing deficit for 1990 at about 1.1 million units. In contrast, the annual supply of formal housing during the 1980s was about 30,000 units per annum maximum (Sundaram 1989:57).

A survey on income distribution in 1981, conducted by the BMRDA, has revealed that the large majority (minimum 75%) of Bombay's population belongs to the lower income groups. Four out of five of such low income households were accommodated in so called *chawls*, mostly dilapidated one-room dwellings of a size not larger than 15 m², with shared facilities such as toilets, staircases and access corridors, and only partially provided with potable water or functioning public latrines. In 1989, nearly 75% of total formal housing stock in Greater Bombay was formed by such *chawls*, whereas the all-India average of one-room tenements amounted to 50% (Sundaram 1989:59-63). The average number of persons per household living in such *shawls* was found to be 6.3, but could go up to as much as 20 to 40 people in one unit. One toilet had, on average, to be shared by six dwellings. Whereas the 1967 Development Plan already used the term 'overcrowding' for these tenements, the situation has worsened ever since. In some districts of Bombay Island the overall density was no less than 20,000 households per hectare (Ibid:61-3).

Beside the formal housing stock of dilapidated structures and new dwellings, another most visible consequence of the scarcity in land and the sky-high property prices are of course the numerous informal self-help structures in slums and squatters, and the number of households living on the pavement. The official 'Census of Slums and Squatter Settlements' already identified nearly 1,700 slum pockets in Greater Bombay in 1976, with a total population of about 3.17 million. At the time, this was about 46% of Bombay's entire population. Nearly four out of five households living in slums, a study by the Tata Institute of Social Sciences revealed, had an income of maximum Rs. 600 per month (Sundaram 1989:64). Over 80% of slum-dwellers were living in the suburbs, and the number of huts per hectare could go up to 300. The slum situation in Bombay did not improve in the years thereafter. On the contrary, between 1961 and 1981 the number of people living in slums increased nearly six times, from around 700,000 to over 4 million (National Commission on Urbanisation, 1987). By 1988 the number was already put at 5.6 million (by the ILO), and for 1995 it was estimated at 68% of the total population (by the Development Institute Bombay).

Those who do not even have the resources to live in a slum are living on the pavement. In 1952 it was estimated that around 20,000 people were living in such conditions, by 1961 the estimate was 62,000 (the majority of which on Bombay Island), and by 1989 the figure had gone up to about 100,000 (Sundaram 1989:24/65).[22] Over 70% of these pavement dwellers had been in Bombay for at least six years, and one in three had come from a slum or *chawl* after eviction by a landlord or by the BMC. These pavement dwellers are, thus, not just migrants that have just moved into the city or nomads, as is commonly believed. About 97% of households living on the pavement have earners, who sustain the city economy in several ways and who contribute to the municipal revenues (Sundaram 1989:65-6).[23] They are thus not the parasites many upper class people believe them to be.

Beside housing, the performance in providing basic urban services has also been pretty poor, even by Indian standards. Quite significant in this respect is that the First Five Year Plan of the federal government allocated 34% of national total investment to housing and urban services, whereas the Seventh Plan reserved only 9% for this purpose (Sundaram 1989:46). Some figures: by 1986, 72% of all households living in formal housing were supplied with water for only 5 to 6 hours per day, and the remaining 28% for only 3 to 5 hours per day. By the late 1980s, not less than 4 million people were getting their water from standpipes, with an average of 270 persons per pipe (Sundaram 1989:59). Obviously, in terms of availability of basic services, much depends on the area and on the socio-economic composition of the occupants. In general, the extended suburbs and low-income settlements are being served worst. Squatters that are not officially recognised (as is mostly the case) may have illegal connections to waterpipes or to the electricity network.[24] In the case of recognised slums, the possibility to implement upgrading programmes is very much dependent on who the land belongs to, and very few slums located on central government land, for example, have enjoyed any kind of improvement.

The sanitation and hygiene record is, unfortunately, not much better than the performance in housing and water supply. According to BMC figures, the number of deaths due to infections and parasitic diseases was during the 1980s an astonishing 40% of total deaths, many of them related to unclean water or bad sanitary conditions (Sundaram 1989:60).[25] Moreover, according to an article by Mukherji in The Independent in 1994, the overall death rate (all causes) is 50% higher in Greater Bombay compared to the adjacent districts of Thana and Raigadh, and 20% higher compared to the entire state of Maharashtra. Compared to the all-India level this is a fairly surprising observation, as death rates all over India are, in general, significantly lower in urban areas than in rural areas (e.g. United Nations 1997:53-4).[26]

Transportation Beside shelter and the provision of basic urban services, transportation has since Independence increasingly been a major problem. The pressure on the road and railway system has become so high, that it has become one of the main bottlenecks for the city economy and a heavy burden on Bombay's millions of daily commuters.

One of the major factors contributing to the transportation problem is of course Bombay's peculiar geographical shape as a narrow peninsular running north to south, with two of the three main labour centres being located in the southern part of the city. Although no statistics can tell what Bombay's deficiency in transportation means in practice, they may give some rough idea. Between 1950 and the 1970s there has been an increase in the number of public transport journeys of around 25% every five years (Harris 1978:34). At present, the two suburban railway lines and public buses together carry about 5 million passengers around every day (Pasricha 1994).[27] The suburban trains carry about a million commuters into the CBD every morning, on trains that are designed to carry 1,750 persons maximum. Most of them however are crammed with 3,500, and in peak hours 4,500 persons; a density rate of over 500 per carriage. Bus transportation, on the other hand, has increasingly suffered from road congestion. Whereas in 1966 Bombay had about 43,000 private vehicles on the road, by 1972 that number had more than doubled to 96,000 (Harris 1978:34). Since then, vehicles have been multiplying thrice as fast as people, at a rate of about 8 to 10% per annum. In 1995, 250,000 private cars and 34,000 taxis were competing for some space on Bombay's clogged roads (Statistical Outline 1995:209).[28] Bombay's public buses carry on average about 1,800 passengers a day each. Although these buses account for over 80% of road travellers, they make up only about 1.5% of all road passenger transport (D'Souza 1986-c). Furthermore, traffic on most roads is also constantly obstructed by other factors than just traffic congestion: illegal squatter settlements, monsoon rains, poor maintenance of roads, insufficient garbage collection, and a very poor general traffic culture.

As with housing, there has been a big gap between growth of traffic and use of public transport on the one hand, and growth in public transport infrastructure and road capacity on the other. Whereas the government policies of the 1970s have mainly been focusing on expanding transport infrastructure, rather than on improving and maximising existing infrastructure, additional supply has been far insufficient. Moreover, too often the policies were designed in favour of huge money-swallowing projects such as fly-overs and freeways, to serve the interests of the privileged 1 or 2% of car owners, instead of the public interest. In more recent years, however, the very priority of expanding such infrastructure has been questioned and it seems that one has finally come to realise that transport infrastructure cannot continue to be expanded endlessly.[29] In terms of general transportation policy, the focus is at present gradually shifting towards privatisation, with legislation being made more flexible and government incentives being introduced to attract private investment.[30] Several other simple as well as more radical solutions, such as the introduction of exclusive bus lanes (as in London) or a system of road pricing and area licensing (as in Singapore) have been put forward but have not yet been implemented.

Debate on future policies

The major symptoms of the dramatic growth of Bombay became increasingly clear during the first ten to fifteen years after Independence. Almost everyone considered Bombay a dying city, a welfare problem, and blamed the mass of migrants moving into the city in thousands as the major evil. Because these migrants were believed to be mainly attracted by the city's job opportunities, the desirability of further economic growth was strongly questioned. The number of people turning a blind eye to Bombay's achievements and economic importance rapidly increased, and arguments in favour of a stop on further urban and economic growth within the city limits gained massive support. Only the government and industry were thinking differently; the government because it was fearing that new investments would be going to other states if not allowed into Bombay, and industry for obvious reasons. Both government and industry, therefore, undertook some mild attempts as to row against the tide, stressing the economic importance of India's major cities, and their role as engines of regional and national economic progress.

It was in this context of disaster scenarios that, from the 1960s, different political echelons started to develop plans and policies for the future. Already in the 1960s, the federal government expressed its concern about the so-called uncontrolled growth of metropolitan cities, but as we have seen previously, this wasn't more than just an expression of concern. The topic 'urban development' was at the federal level not even a separate policy chapter, as it was an integral part of the chapter on housing. Until the 1980s the main line in the Planning Commission's arguments was simply that the metropolitan cities had become too large, and that in cities of the size of Bombay and Calcutta urbanisation had reached the limits of saturation. Therefore, further urban and economic growth was to be decentralised as much as possible, and a licensing policy, the 'Industries Development and Regulation Act', was introduced. New industries had to be stopped from establishing themselves on Salsette Island, and existing industries and certain parts of other sectors had to be motivated to shift out of Bombay.

The response of the Government of Maharashtra to the problems of Bombay was perfectly in line with the ideas and attitudes at the federal level. From the early post-colonial years onwards, the state's urban development and housing policies were mainly focused on shelter issues, such as the use of land, public housing, renovation of dilapidated stock on Bombay Island, slum clearance and improvement, and helping the municipal authorities with investments in water supply and sewerage or in public transport. On the non-housing front, the state's policy, in the initial phase, consisted of not really more than industrial location and dispersal and of master planning. At the same time, however, several study groups and special committees were established to look into the matter of excessive metropolitan growth.

Study groups and special committees, and the roots of the twin-city A first modest attempt was made soon after Independence by the Development Committee under chairmanship of Mayer and Modak, which resulted in a first Master Plan ('Master Plan in Outline', 1947) to regulate and plan Bombay's economy. Some points that were recommended by Mayer and Modak were the inclusion of the Salsette towns into the Bombay administrative area and the banning of industries: the light industries to the suburbs, and the heavy industries to the northeastern area on the mainland. Another recommendation was the setting up of satellite towns on the mainland (Honkalse 1992:57). Although their proposal on inclusion of suburban areas was to a large extent executed in 1950 and 1956, the proposal on satellite towns was not given due consideration in political circles. For some, however, the Mayer and Modak report was considered the root of the New Bombay idea (Honkalse 1992, Ray 1989).

Earlier, however, there had already been two studies in which development of the area opposite the harbour on the mainland had briefly been mentioned. One was by a state government committee that had come up with the suggestion,[31] the other one was a feasibility study by the Indian Railways to extend one of the railway lines onto the mainland towards Nerul and Panvel in current New Bombay.[32] However, neither of these studies was accepted, and the proposal of developing the area opposite Bombay was thus indefinitely postponed.

In the meantime, some restrictive measures, such as the 'Bombay Building Works Restrictions Act' of 1949, introducing restrictions on industrial location, were taken at the municipal level, and Bombay's municipal boundaries were being extended to include five municipalities in the suburbs. Consequently, the administrative name became 'Municipal Corporation of Greater Bombay', which soon passed another restrictive bill on industrial location on the island, viz. the 'Bombay Municipal Corporation Act' of 1951, introducing zoning on the island to discourage new industries, and to encourage a shift of existing plants to the extended suburbs. The passage of the Bombay Town Planning Act in 1954 empowered the MCGB to prepare Master Plans, to define land uses and to operate development controls. In 1956 another 26 villages were being included into the Greater Bombay area.

Two years later another study group was formed under the chairmanship of S.G. Barve. The group produced the so called Barve Report ('Report of the Study Group on Greater Bombay', 1959) which also suggested decentralisation of industry and housing, a ban on establishment of most new industries on Bombay Island (supported by the federal licensing policy), the building of a bridge eastwards connecting Bombay with the mainland, and development of a township opposite the harbour to take industry dispersed from the city. The Barve Report, however, made a much stronger argument for development of the area opposite the port than any other document or study so far, as it stated that 'the error of history must be corrected and the handicap of geography must be overcome by developing a city on the mainland-side of the Bay'. Such a development, the report continued, 'would have multiple effects: it would relieve the pressure on the railways and the roadways... , it would provide room for the development of a port on the other side of the Bay..., and it would open up the possibility of the development of a new township, which would in time not only draw away from the Island of Bombay the overflow of the industrial units and residential colonies, but also enable the City to develop in an orderly manner' (Correa et al. 1967).

To my feeling, the Barve Report must be considered the real roots of the new town. Furthermore, the report included some other new elements, such as the suggestion of a new commercial centre at Bandra-Kurla (the most southern area of the Bombay Suburbs), and relocation of the main administrative offices on Bombay Island. The Group also felt the need for an integrating planning and development authority at the state level. Some of its recommendations, such as the building of the bridge eastwards, some suggestions concerning industrial location, and the Bandra-Kurla Complex, were accepted by the state government.

In 1964, the BMC produced a proper 'Development Plan for Greater Bombay'. It decided, among many other things (including large-scale investment in housing and urban services and their financing), on a dispersal of population to the suburban areas, and on a decentralisation of specific industries and commerce from the island (Verma:25).

However, it was increasingly realised by the government that a plan that confined itself to the limits of the BMC would not be entirely adequate as a solution to the complex problems of Bombay. Therefore, the state government, in March 1965, appointed another important planning committee headed by the distinguished economics professor D.R. Gadgil, in order to prepare an extensive regional plan for metropolitan Bombay and its periphery. The Gadgil Committee feared that Bombay was gradually falling to pieces, and policy-makers were to take 'vigorous action so that the threatened increase in population... does not... take place', and the population would stabilise 'at a figure not higher than 25 lakhs' (Gadgil Report:74-75). Consequently, much more drastic action was considered necessary, with a nearly complete ban on industrial activities on the island, location of all new offices in the Bandra-Kurla area, and in relation to suburban areas 'to reduce commuting and establish relationship between residence and work place, schools, shops and other amenities' (Ibid:73-74). Furthermore, the committee also identified the Greater Bombay Region and proposed the development of the mainland 'by creating a multinucleated metropolitan region... making it economic to operate providing proper relation between working and living', as well as 'a development authority with adequate legal powers' (Ibid.). Again, these recommendations were selectively accepted by the state government.

Meanwhile, the state government had also established a number of state agencies to specifically support industrial decentralisation and development over the entire state, a move which was quite unique for India in that time. In 1962 the Maharashtra Industrial Development Corporation Ltd. (MIDC) was formed to assist industrial development over the entire region of Maharashtra by developing industrial estates and providing the necessary infrastructure. In 1965 the State Industrial Investment Corporation of Maharashtra Ltd. (SICOM) was created to administer and publicise a package of incentives (excluding the areas around Greater Bombay which were promoted by the MIDC). One of the areas that were reserved for MIDC industrial estates is located in current New Bombay, of which the Gadgil Report stated that 'A much larger urban development than that of a satellite town is bound to take place in this area. Detailed planning of this area should be undertaken immediately' (Gadgil Report:75).

Although such definitions as 'planned industrial decentralisation', 'employment dispersal' or 'development of backward regions' were frequently used in almost every official planning document of the state government or the BMC since the early 1950s, the reality was quite different. Whereas controls over existing industries were only marginally created, controls over new investments in industry in the Bombay region were very strictly defined, but largely ignored in practice. Since the early 1960s, an increasing number of incentives to induce firms to move to more backward districts in the state have been introduced by the government or state government agencies (MIDC, SICOM).[33] However, the game of politics, and more specifically the struggle between different states, and between different regions within a state, for scarce financial resources and for public and private investments has caused many distortions and a very large part of total investments is still absorbed by Greater Bombay and its periphery. In Maharashtra for example, by 1965 the MIDC had created twenty industrial estates, but half of these were located in the area that was later to become the Bombay Metropolitan Region, BMR (Harris 1978:52). In more recent years, the investment pattern in Maharashtra has showed a similar pattern, with the large majority of large-scale investments done in and around Greater Bombay (see table 6.3).

At the inter-state level, there were numerous attempts of other states to exploit the disadvantages of Maharashtra's industrial policy by enticing firms to invest their capital in metropolitan regions outside Maharashtra. Consequently, the state government increasingly granted exceptional permissions to firms threatening to leave the state, and certain industrial policies and definitions were openly relaxed so that, in the end, few formal controls were left to keep new industries out, or to prevent existing industries from expanding or from renewing their licenses. It is evident, therefore, that one of the main factors in this process of erosion of the state's industrial decentralisation policy has been the lack of a properly outlined national policy for the entire country. The absence of such a policy has clearly been an open invitation for inter-state rivalry for industrial investment and location.[34]

In 1967, in the same period that the Maharashtra Regional Town Planning Act was passed, the Bombay Metropolitan Region was defined (largely as a result of the Gadgil Committee's recommendations) and notified by government on 8 June 1967. Furthermore, a BMR Planning Board (BMRPB) was constituted on 31 July 1967.[35] The Boards' only task was to develop a regional plan for the next twenty years. Although the planning reports of the previous period had provided numerous proposals, not much had been put to practice. The reasons for the inertia were numerous: lack of political vision and insight in long-term economic processes, overlapping planning and development responsibilities, a gap between planning potential and implementation and between development and maintenance, financial constraints, political interests, and so on. Hence the importance of the BMRPB as a co-ordinating planning agency for the larger Bombay region.

Around approximately the same time as the establishment of the BMR, another very important but non-official document was published, based on a symposium and a number of privately organised seminars. It was the period in which the enthusiasts of the twin-city concept, as it had been outlined in the Barve Report eight years earlier, were pushing hard on government agencies, especially on the BMRPB, to accept the twin-city proposal. The document was published as the 'MARG Piece' (Modern Architects Research Group),[36] and was strongly in favour of development of a twin-city opposite Bombay, instead of further development on the Salsette Peninsula or development of new satellite towns on the mainland. The authors argued that strong 'counter-magnets' were needed 'if we are to draw away a sizeable percentage of the hordes *(sic)* of immigrants who will otherwise descend upon the existing cities in the next 35 years' (MARG Piece:31).[37] The question was put forward which alternative, given

Bombay's unusual structural pattern, would serve the future of the city best. The choice was between either an acceptance of the existing centre of activity as the continuing focus for future development and an attempt to deal with the increasing pressures, or the creation of alternative centres of activity to serve as counter-magnets to draw the pressures away. In view of the expected further increase in population and economic activity, the second alternative was considered the only one possible.

In order to create counter-magnets, different techniques are possible. One of these, the creation of a number of satellite towns, was for several reasons not considered suitable,[38] and thus the ultimate choice of the government had to be 'a single major urban centre... on the mainland directly opposite Bombay, of equal prestige and importance', so that it could 'develop into an area as large as the older city' (MARG Piece:36). Furthermore, the MARG authors proposed a shift of government offices to the new city, which thus would become the new capital of Maharashtra, and overall the twin-city was to become 'a new self-sufficient city, whose continual contact and exchange with the older city would be primarily for social, cultural and commercial reasons and not as a 'dormitory' area feeding the already overcrowded island' (MARG Piece:43). At the same time, the authors levelled criticism against alternative proposals, such as the one to develop the Bandra-Kurla area in the Bombay Suburbs as a major second new business centre, fearing probably that this would ruin the chances of the twin-city. Financially, it was agreed that this project would not cost much more than implementing the municipal Master Plan of 1964 in which, according to the authors, nearly a quarter of the total development cost would be for land acquisition. All in all, the authors concluded that 'the development costs will be practically identical for either location, and the only additional costs for the alternative plan would be for land reclamation and for the major bridges and highways' (MARG Piece:48).

The MARG Piece has evoked criticism from several corners, with the sharpest reaction coming from H.S. Verma in his book 'Bombay, New Bombay and Metropolitan Region. Growth Process and Planning Lessons' (1985). In this book Verma attempted to uncover the real motives of the key actors involved in each major decision episode on the way to the acceptance of the twin-city proposal by the state government. He argued that the proposal for a twin-city was a simple continuation of what he called the 'systematic methodology of growth in Greater Bombay', which in his view had always been actively supported and influenced by the business class. In his opinion, New Bombay was a plan 'by the business class, and for the business class'.

Formal proposal for development of a 'New Bombay' Between 1967 and 1969, the BMRPB, whether or not influenced by pressures from business circles (as Verma believed) or essays like the MARG Piece, started to discuss and to review previous proposals for development of Greater Bombay, especially the 1964 Development Plan. The Board changed its views quite drastically on several crucial development issues. These views were formulated in a draft report, which was published for public comments on 27 January 1970. The suggestions and objections received were scrutinised and the plan finalised in October of the same year, after which it was sent to the state government for approval.

In general, the BMRPB considered all previous proposals as mere palliatives. The Planning Board agreed with the arguments in the MARG Piece, that in view of the predicted population growth for Greater Bombay (at the time estimated to be 9.81 million by 1991 as compared to the 5.4 million of the 1961 census), the metropolis would not be able to accommodate another 3 to 4 million people or more. Moreover, the BMC's Development Plan, which had been prepared to meet the needs of a population of only 7 million, was already expected to cost about Rs 1000 crores (at 1970 prices), and was therefore in itself 'largely unrealisable due to the non-availability of financial resources of the required order' (BMR Regional Plan:93). Decentralisation and dispersal, the Regional Plan concluded, was therefore to be the central theme, and the main question was which pattern of future regional development was to be most suitable for that purpose.

The BMRPB identified four alternative patterns for future regional structure: (a) an internal restructuring of the metropolis without any new towns; (b) a multi-town structure or a series of medium-sized new towns; (c) linear urban corridors; and (d) a single large counter-magnet to constitute a new metro centre. The first three patterns of regional structure were either considered not completely satisfactory or not possible. Internal restructuring without any efforts for decentralisation and dispersal away from Salsette was considered as a dead-end solution after some years. The idea of a multi-nucleated structure of medium-sized towns, initially recommended by the Gadgil Committee, 'would not by itself be sufficient to relieve congestion in Bombay' because 'The population absorption capacity of a series of small or medium-sized new towns would be necessarily limited ... and not be capable of attracting the tertiary sector activities away from Bombay' (BMR Regional Plan:104). The latter would only be possible in a metropolitan setting and not in several small new towns dispersed in the region. The third alternative of linear urban corridors in the wider Bombay region, already developing spontaneously along the major road and railway lines, was also considered as far insufficient in the context of Bombay as 'they would merely be linear extensions emanating from the existing centre and would have no effect in evolving a new centre or focus for the large scale development of the tertiary sector, and therefore have no worthwile effect on easing the trend of growing congestion at Bombay's core'. Moreover, they would add to the 'pressure that they inevitably create on the linear communication' (ibid.).

Only the fourth alternative, possibly in combination with elements of the other three, was believed to have the potential to be successful. That fourth alternative, a new metro-centre, was to consist of 'the development of a twin metropolis on the mainland which would provide the fabric necessary to attract employment-intensive tertiary activities and also provide social, cultural and other amenities for the surrounding area' (ibid:104). Any solution which would fail in arresting the continued rapid growth of the tertiary sector in South Bombay (considered as largely responsible for the deteriorating situation on the island) was to provide only marginal relief according to the Board. 'To be really successful in re-orienting growth of Bombay', the BMRPB argued, 'new development will have to be of such an order and of such character as will be able to induce the sophisticated tertiary sector to locate itself away from the congested southern tip' (ibid.). In relation to this, the BMRPB argued that 'Government is at the key position in this sector' and shifting the major government offices from the island to the new town was thus to be an integral part of the strategy and was considered crucial if any growth in the new town was to be achieved within a reasonable time (Ibid:109/10-164).

Another factor that favoured the idea of a new metro-city on the mainland, was the economically stimulating impact such a metro would have on the relatively backward adjacent areas inland. Where exactly on the mainland this new twin-city was to be located was fairly easy to decide. The area exactly opposite Greater Bombay was considered as ideally suited in many ways, since it already accommodated the Trans-Thana creek bridge and the new expressway to Poona, the new docks on the Nhava-Sheva shores and the Diva-Panvel and Panvel-Uran links of the railway. These realisations were considered as containing within themselves the roots of large growth and, therefore, the Planning Board concluded that 'it only stands to reason that the potential created by this investment is exploited to the full' (ibidem:105). Moreover, in view of the existence of industrial zones already established by the MIDC in the area (the Thana-Belapur industrial belt and the Taloja industrial estate), and further likely development of port-oriented industries around the port, enlarged growth potential was expected to be created in this area anyway. The proximity of Greater Bombay, although separated by a wide creek, would 'ensure its growth as an independent city' (ibid.) and the old overcrowded north-south transportation arteries would be relieved by a new east-west transportation axis.[39] Ultimately, the Board considered no single solution as adequate to take care of the development problems of the metropolitan region, and the future pattern was therefore to be a combination of the four alternatives, with the single most important proposal clearly being the development of a twin-metropolis. After including certain modifications, the government accepted this Regional Plan, and the date for the plan to come into force was fixed at 16 August 1973.

Two things are worth noting. First, that the chapter on the twin-city in the final draft of the Regional Plan was almost an exact copy of what the authors of the MARG Piece had suggested earlier. And secondly, that the BMRPB, being a state agency, had gone a long way from the initial state policy of industrial decentralisation and dispersal, as the area designated for development of the twin-city was just a stone's throw from Bombay. These observations, however, do not alter the fact that, in the given circumstances and within the existing geographical and urban context, the idea of planning and developing the area across the harbour was probably the best possible long-term solution for the city of Bombay.

Conclusion

The historical growth process of the city of Bombay has mainly been the result of a number of economic and political factors, both during the colonial period and after Independence. These factors have caused dramatic changes in the demographic and spatial urban patterns, and have beside wealth and progress generated a number of infrastructural inadequacies and overall urban congestion. The discussions on future policies, which followed in the 1950s and 1960s, led to the ultimate acceptance of the so-called New Bombay Plan and the creation of a twin-city on the mainland. How the planning and development of this new area was conceived in the early 1970s, and what has come out of it after twenty-five years of project implementation, will be the major focus of part II of the book.

Notes

[1] Estimate of the Portuguese church.

[2] Surat is about 300 kms north of Bombay. In those days it was the largest port on the subcontinent.

[3] The name *Salsette* seems to be derived from the Marathi *sashti*, which refers to the *sixty villages* which were spread all over Salsette Island.

[4] The Bombay Presidency was a totally different area from the current state of Maharashtra. It stretched out over a huge area including cities such as Karachi (Pakistan), Ahmadabad (Gujarat state) and parts of current Karnataka state in South India.

[5] The most important factors in the instability were the British annexation of Oudh, the use of the *doctrine of lapse*, the structural adjustments in the Indian countryside, and the introduction of new Enfield rifles in the army.

[6] The BMC has always been a pre-eminent municipal body in the Indian local government scene. Created by separate provincial legislation it was, until recently, the only municipal corporation in the country which could not be superseded by the state government. Obviously, this has always been of significant importance in terms of political autonomy and independence. Its privileged position came to an end in 1984 through an amendment to the Act by the state legislature, followed by the appointment of an administrator in lieu of the elected corporation.

[7] Horvath's model describes the colonial city's stratification system, with the colonial *masters* forming the elite, an intermediate group consisting of different races and immigrants coming from outside (but from another country than the coloniser), and the local people at the bottom of the system, some of whom would be involved in the lower colonial administration.

[8] As compared to the more recent industrial suburbanisation.

[9] India was in that time the second largest exporter of cotton in the world.

[10] This *Bombay Suburban District* was only a fraction of what is today known as Greater Bombay.

[11] Since the turn of the century it was the *Bombay Electrical Supply and Transport Undertaking* (BEST) which was responsible for the entire urban transport sector.

[12] A special census was organised in 1906 after, and because of the plague epidemic.

[13] The ratio in the Ratnagiri district, a poor and infertile area 250 miles south of Bombay, from where a large number of these migrants came, was in the same period 84 males per 100 females. The ratio of males aged between 15 and 45 years old, as compared to the total number of males, was 37% in 1881 and 48% (nearly half!) by 1921.

[14] A survey between 1921 and 1923 concluded that not less than 97% of working class people were living in one-room chawls (Harris 1978:11)

[15] In 1951, 35% of the total work force was employed in industry (compared to 25% in trade and services), and 41% by 1971 (compared to 24% in trade and services).

[16] The percentage of factory workers in the Bombay region (Greater Bombay + Thana District) as compared to the total for Maharashtra was 49.1% in 1991.

[17] By the early 1970s, the amount of goods passing through Bombay harbour was by far the highest in the country, not less than 40% higher than in Calcutta (Harris 1978:15). Exports accounted for 3.7 million tons (mainly engineering products, textiles and minerals) and imports for 12.4 million tons (increasingly petrochemical products), more than half of which stayed within the larger Bombay region. The over-concentration of economic activities in and around Bombay seems, however, to have decreased in the period thereafter. Whereas the percentage of factory workers in the Greater Bombay and Thana region together, as compared to the entire state of Maharashtra, was still 57.09% in 1984, it went down to 53.83% in 1987 and to 49.1% in 1990 (*Selected Indicators for Districts in Maharashtra, 1984-85/1987-88 and 1990-91*).

[18] The kind of workers that are to be found in Bombay's informal sector include tailors, potters, shoe makers, glass makers, gem cutters and other such handicraft workers, street vendors, repairers, rag-pickers, recycling workers, etc. They all have set up a small business in Bombay or in the rural periphery, and produce almost exclusively for the Bombay market.

[19] Whereas in the most recent census of 1991 the overall population on Bombay Island had actually decreased with 3.5%, there has been a clear distinction in spatial distribution of this negative growth. The population increased in the most southern ward (ward A) with 16%, and in the two most northern wards (F and G), the districts nearest to the Bombay Suburbs, with 4% and 2% respectively. In the other four wards (B,C,D and E) population growth was negative and went down with 21% on average. The main reason for this peculiar pattern is probably the enormous increase in property rates in the areas close to the CBD. The exceptional population increase in ward A can be explained by the constantly increasing supply of high class residential space on reclaimed land in that ward.

[20] J.B. D'Souza is a retired IAS-officer who is still considered as one of the most distinct urban specialists in Bombay and in India. He has held several top positions in the local and state administrations, one of which was the function of first Managing Director of the development authority in New Bombay.

---

[21] In one suburban district the population increased with 1600% during 1961-71, while the number of workers increased with 1150% (Harris 1978:21).

[22] These are estimates, since these pavement dwellers have always been left out of the official census.

[23] Pavement dwellers are employed in typical low class jobs such as self-employed occupations, domestic service, rag picking etc., although this segment of the population also includes artists, sculptors, composers and even municipal workers or railway and other government staff.

[24] The official policy on slums makes little sense as it only tolerates slums that have existed before the slum census of 1976. Newer slums are constantly threatened to be evicted.

[25] As has been said in previous chapters, it should be noted that Bombay is not a unique case in all this. F. Engels for example used to say pretty much the same things about Manchester in the 1830s and 1840s, and London also exhibited appaling housing and health problems up to the 1870s.

[26] At the all-India level death rates [per 1,000 population] were in 1971 16.4 (rural) and 9.7 (urban), in 1981 13.7 (rural) and 7.8 (urban), and in 1991 10.6 (rural) and 7.1 (urban) (Statistical Outline 1994-95).

[27] Dr. Pasricha is Bombay's Inspector General of Police for City Transport, and the first person in India who obtained a Ph.D. in traffic management.

[28] Compared to other major Indian cities, Bombay is having considerably more private cars than Calcutta but less than Delhi. The number of taxis is about twice as much as in the second 'taxi sahar' Calcutta.

[29] Nevertheless is there another example in the making. Since the suburban railways have, at present, again reached a saturation point, with each train having nine carriages and (in theory) one train passing a station every three minutes, plans are now circulating to simply add three carriages to the suburban trains and let them run every two and a half minutes (from a discussion with V.K. Phatak, the Chief Planner of the BMRDA). Firstly, another extension of trains is not that simple, since all the platforms will have to be extended over a distance of three carriages, and secondly, if there really would be a train every three minutes there would not be such a problem.

[30] As Pasricha has pointed out, the cost for expansion and maintenance of Maharashtra's highways alone, would go into thousands of crores and the government on its own simply cannot handle the situation anymore (Panjwani 1994).

[31] The Bombay City and Suburbs Postwar Development Committee, which produced the *Preliminary Report* (Development of Suburbs and Town Planning, Housing and Traffic Panels, Gov of Bombay, n.d. but probably 1946)

[32] MTP Railways, 1980, *Techno-Economic Feasibility Report on Extension of Railway Line from Mankhurd to Belapur*, Bombay ; Panvel was at that time an already considerably developed municipality (see further).

[33] Incentives were for example the provision of excellent infrastructure, a lower leasing price for industrial plots further away from Bombay, assistance in obtaining concessional finance from the MSFC (the state's Finance Corporation), and certain tax incentives.

[34] In all this, Bombay was certainly not a unique case in the world. In Tokyo for example, in the 1950s and 1960s, numerous industrial location controls had been introduced, but most of them were ignored, or allowed to be ignored.

[35] The BMR as notified by the state government was delimited as follows : 'in the west by the Arabian Sea, in the east by the eastern administrative limits of Kalyan and Bhiwandi tehsils and foothills of Sahyadri in Karjat tehsil, in the north by the Tansa River, and in the south by the Patalganga River'.

[36] Charles M. Correa, Pravina Mehta and Shirish B. Patel, *Planning for Bombay. Patterns of Growth. The Twin-city. Current Proposals* (n.d., probably May 1967).

[37] This was probably the first time that the word 'counter-magnet' was used.

[38] To serve their purpose, the townships, according to the authors, had to be large (about 1 million people), they had to be near and located much more in the form of a half-ring around Bombay. It would be these satellite towns which had to absorb most of the future growth. Each town would have to be equipped with all the secondary services of a metropolitan area. If only marginal services were to be provided, the authors argued, the satellite towns would inevitably grow as shanty towns, housing only those who have industrial employment in the immediate neighbourhood. The pressures on Bombay would not be relieved. Finally, the total cost of several new satellite towns would be exorbitant (MARG Piece:36).

[39] The plan did not mention, however, that this new east-west corridor would connect to the north-south axis somewhere in the suburbs, and therefore add to the pressure on the existing road and railway lines going into the CBD of South Bombay.

PART II

BIRTH OF NEW BOMBAY AND IMPLEMENTATION OF THE NEW BOMBAY PROJECT

5 Birth of New Bombay:
Policies and Strategies

Figure 5.1: Greater Bombay and New Bombay (shaded)

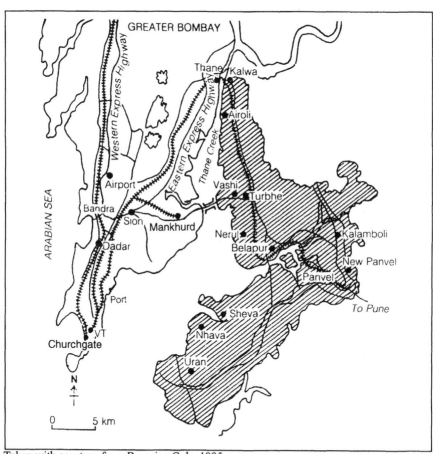

Taken with courtesy from Banerjee-Guha 1995

Introduction

In the previous chapter we have seen how after more than a decade of debate and discussion on the future development strategy for Bombay, the original Regional Plan for the city was, in the course of the period 1972-73, drastically changed in favour of a single new independent town of metropolitan size. In this next chapter we will first discuss the creation of a planning and development authority for the new town and its major tasks and functions, as well as the proposal for a future administrative structure. Secondly, we will consider the physical and socio-economic features of the area at the time of notification in the early 1970s. The final and major part of this chapter will discuss the planning policies and development strategies that were put forward in the frequently adjusted New Bombay Development Plan. This plan was the first real creation of CIDCO's selected planning team, and is an expression of the general ideas on planning and development that were prevailing within the development agency at the time. It gives detailed information on general planning concepts, land use pattern, broad social and physical infrastructure development, on ways to rehabilitate and integrate the local population, and on the mechanism in which the project was to be made financially sound.

Establishment of a planning authority and creation of an administrative structure

Unusually quickly after the Regional Plan was finalised in August 1973, the state government accepted the recommendations of the Planning Board to set up a new town, or 'metro-centre' as it was called, on the mainland.[1] An area of 344 km² was designated, covering 95 villages from Thana, Panvel and Uran *tehsils*, spread over the two Districts of Thana and Raigad.[2] Privately owned land was notified for acquisition under the Land Acquisition Act 1894, and the new area was named 'New Bombay'.[3] CIDCO, the City and Industrial Development Corporation of Maharashtra Ltd., which was formed on 17 March 1970 as (initially) a subsidiary of SICOM and was registered under the Indian Companies Act 1956, is a limited company which is wholly controlled by the Government of Maharashtra.[4] In March 1971 CIDCO was designated the New Town Development Authority (NTDA) for the New Bombay project, after which it became fully responsible for planning and developing the area.[5]

Immediately thereafter a Planning Team of eight experts in various fields was set up to help CIDCO in its planning effort.[6]

As the New Town Development Authority, CIDCO's functions have been planning, implementation, maintenance and administration of New Bombay, and of other areas which have been entrusted to it by the state government afterwards.[7] Thus, as long as no local municipal body would be established for the New Bombay area, CIDCO would also be fully responsible for providing civic services to the residents of the new town.

Only in December 1991 the state government would constitute such a municipal corporation for New Bombay, viz. the Navi Mumbai Municipal Corporation (NMMC), which became operative with effect from 1 January 1992. Due to legal and practical problems, the actual shift of administrative responsibilities from CIDCO to this first democratically elected NMMC has only been executed very recently and only partially, and the overall situation of administrative jurisdiction has been, and still is, very confusing to all parties involved.[8] This new municipal corporation includes 44 villages, which were formerly administered by 28 village 'panchayats' (local councils at the village level), although only 29 of these villages (viz. the villages falling within the Thana Tehsil of Thana District) fall within the New Bombay project area. The remaining 15 villages under the jurisdiction of the NMMC fall outside the New Bombay project area, east of the Parsik Hills.

The development procedure was to be such, that as soon as CIDCO, being the NTDA, had completed the development of a new 'node' (as the new townships of New Bombay, of which there were to be twenty, were called in the NBDP), jurisdiction over that node would be transferred to the NMMC, which would then become the future administrative and planning authority for that node. However, this arrangement was only confirmed officially by a government notification of 16 December 1994. By January 1995, five newly developed nodes within New Bombay, viz. Vashi, Nerul, Belapur, Koparkhairane and Airoli, had been shifted to the NMMC. As development in the southern part of New Bombay has only recently started, CIDCO is still fully responsible for the planning and development of this area (the municipal towns of Panvel and Uran not included, see below). According to A.B. Sapre, additional Chief Planner of CIDCO, the entire New Bombay region, once fully developed, will probably be split under three municipal corporations. One big Panvel MC, one big Uran MC, and one big New Bombay MC (NMMC). Although it was initially the plan to create one large municipal corporation for the entire area of New Bombay, it is now expected that from the moment that the southern nodes will be fully developed as well, it will probably be either Uran or Panvel who will take over the administrative and planning responsibility over these areas.[9]

The New Bombay area at the time of notification

What did the area look like before the NTDA moved in in the early 1970s? Apart from many small villages, there were two municipalities with a certain history, viz. Panvel and Uran.[10] The town of Panvel had a population of 14,861 in the 1951 census, Uran had a population of 8,672 (Maharashtra State Gazetteers 1964:209,624). These two towns, covering a total area of about 14.25 km², were administered by two municipal councils, and were therefore excluded from CIDCO's jurisdiction. The same was true for the original villages or 'gaothans', in which CIDCO would not undertake any development work.[11] Apart from these villages and two towns, the areas under MIDC, the MSEB (Maharashtra State Electricity Board), and the Defence Department were also part of New Bombay's total project area of 344 km² but were not included in the actual development project. Therefore, the net area of the project was about 50 km² less, to be around 294 km² : 56.3% (166 km²) of this was private land, 34.5% (101 km²) was government land, and 9.2% (27 km²) was salt pan land. Nearly 15% of total land to be acquired was non-usable land that consisted of forest, green zone, *khajan* (low lying land), and scattered pieces.

The New Bombay area, as it was outlined and notified by the state government, can be described as an area consisting of two halves which lie more or less exactly opposite Greater Bombay and roughly run over the same distance north to south, with the Panvel creek and the Vaghivali island in the centre (see page 109). One half runs north to south over a distance of about 25 kilometres and includes the MIDC areas of Taloja and the Thana-Belapur industrial belt; the other half runs north-east to south-west starting from the Taloja industrial estate over a distance of about 30 kilometres and includes the two municipal towns of Panvel and Uran, and the two former islands of Nhava and Sheva.

New Bombay's planning and development boundaries were defined following to a great extent the natural features of the area. In the west and south-west, the boundary is formed by the water of the Thana creek; in the north it extends the Greater Bombay's northern boundary onto the mainland where it attaches the town of Thana; in the north-east the boundary is defined by the Parsik Hills and by the MIDC industrial estate of Taloja; in the eastern part of the central area by the administrative boundaries of villages between Taloja and Panvel, along the foot of the Adai Hills; and in the south by the villages of Kalundre and Chanje, and over some distance along the Karanja creek.

The development potential for the area at the time consisted of the MIDC industrial estates in the Thana-Belapur industrial belt and the Taloja industrial estate; the planned concept of a new and modern second port at Nhava Sheva, and the ONGC offshore base at Nhava in the south; the commissioning of a new bridge across the Thana Creek linking the new area with the Bombay Suburbs; and the existence of the two older municipal towns of Panvel and Uran, and a number of major highways already running through the area.

At the time of notification 24,878 households were living in the project area, 19,400 (or 117,000 persons) of which were spread over the 95 villages included in the notified area of New Bombay (CIDCO 1973:16/103). The large majority of locals were either agricultural labourers, fishermen or salt pan workers, with the big majority involved in agriculture and owning their own small plot of land. For many, agriculture was a temporary activity limited to the four monsoon months between June and September, and they therefore combined agriculture with fishing or other activities.

So far as industry is concerned, the 900 ha. of land reserved for the Taloja industrial estate were still almost completely undeveloped, and the Thana-Belapur industrial belt (TBIA) was the only fairly developed industrial area in the region before it was officially notified as such. The TBIA was covering a total area of 2,105 ha. The first industrial unit in the belt had already been established in 1951, but until 1965 there had hardly been any industrial growth. At that moment, however, the MIDC became actively involved in stimulating industrial development in the area, and following the establishment of the 'Chemicals and Fibres of India' plant in 1965, the number of industrial units increased rapidly to 44 by 1971. Nevertheless, together these plants occupied not more than 9% of total area available in the TBIA (Bhattacharya 1971:28). The head-offices of most plants were based in Greater Bombay. The total work force employed in the belt was only 15,732 in 1971 (Bhattacharya 1971:11), wages were relatively high compared to industrial wages in Bombay and elsewhere, and capital investment of Rs. 109,000 per person employed was very high as compared to an average of about Rs. 23,000 in Greater Bombay and Rs. 20,000 in Maharashtra (CIDCO 1973:90). This indicates the presence of capital-intensive industries such as chemical and petro-chemical industries, alongside a number of engineering units. Land was relatively very cheap, and economically necessary facilities such as water, power supply and roads had been provided by the MIDC. Other essential facilities however, such as housing, drainage, public transport, telephone/telegraph connections, recreation, banking and medical facilities were chronically lacking. Consequently, the majority of industrial workers were living in Greater Bombay or Thana and commuted daily by company bus or public bus to New Bombay. Only 23% of industrial employees came from nearby villages, and 3% were accommodated by the employer close to the work place (Bhattacharya 1971:14).

Already at the time of notification, the area was having a serious problem with industrial pollution, because the Parsik Hills, with peaks up to 235 metres above sea level, have always been a natural obstruction in the east for gases emitted from both the industries in the Bombay Suburbs, drifting with the south-western winds across the Thana creek, as well as from the industries that were already located in the Thana-Belapur industrial area itself. Another special feature of the area was that out of a total of about 34,000 ha for the entire project area, about 20% was low-lying land and therefore prone to tidal action with saline water entering these areas from time to time. Part of those lands were being used for salt making, but most of them were marsh lands. Consequently, although not all of these lands were to be drained, large reclamation schemes were bound to become an important part of CIDCO's activities in the first couple of years.[12]

The New Bombay Draft Development Plan : programmes and policies

In October 1973 CIDCO published its New Bombay Draft Development Plan (NBDP) for public comments and suggestions, as was required by the Maharashtra Regional and Town Planning Act 1966, and submitted it to the state government for approval in October 1975. It was finally sanctioned on 18 August 1979, but was modified several times afterwards by the CIDCO Board of Directors. Preparation of the NBDP took more than two years, as it was the result of a long series of discussions on planning and development strategies to be followed. In every planning step that was taken, the NTDA officials were compelled to focus on the five broad objectives for the project put forward by the state government (CIDCO 1973:10):

(1) to reduce population growth in Greater Bombay by creating an attractive urban centre on the mainland across the harbour, which would (a) absorb migrants who would otherwise go to Bombay, and (b) attract some of Bombay's present population, in order to keep the overall population of Greater Bombay within manageable limits;

(2) to support state-wide industrial location policies which would eventually lead to an efficient and rational distribution of industries over the state and to a balanced development of urban centres in the hinterland;

(3) to provide social and physical services, which would raise living standards and reduce disparities in the amenities available to different sections of the population;

(4) to provide an environment which would permit the citizens of the new town to live fuller and richer lives, free as far as possible of the physical and social tensions associated with urban living; and

(5) to provide training and all possible facilities to the existing local population in the project area, in order to enable them to adapt to the new urban setting and to participate fully and actively in the economic and social life of the new city.

Furthermore, in the initial phase the NTDA had also to undertake a series of preliminary studies in order to gather data on all sorts of physical, economic and social aspects, as a clear understanding of the context in which planning decisions were to be taken was obviously of crucial importance. Finally, CIDCO's policies were also to be brought in line with the programmes and policies of other state and central agencies that were going to be active in the project area. Ultimately, after two years, all these inputs were drawn together in the formulation of an overall planning and development concept, as outlined in the Development Plan.

Given these state objectives, CIDCO defined its future principal activities as :

(a) to develop land and provide the required physical infrastructure such as roads, drainage, water supply, sewerage, street lighting, landscaping, etc;

(b) to build as many houses and community centres, shopping centres, parks, playgrounds, bus stations, etc. to meet the day to day needs of the population as well as for a faster take-off of new growth areas, and to make available developed plots at affordable prices to people to construct houses;

(c) to promote growth of commercial activities, wholesale market activities, warehousing, transport, and office activities in order to evolve expeditiously a sound economic base for a self-sustained growth, achieving at the same time a process of relieving congestion in the old city of Bombay; and

(d) to involve agencies in the development of public transport, both road and rail, and of telecommunication (CIDCO 1986, 1989, 1992).

Nodal development concept

The NBDP envisaged a policentric pattern, or 'nodal settlement pattern' to be build up along major transport corridors.[13] Each node was planned to ultimately have a population of 50,000 to 200,000 (CIDCO 1973:17-25). In subsequent planning documents the minimum was brought to 100,000 per node, with each node to cover an area of 400 to 800 ha of land, and to be subdivided into sectors (CIDCO 1986, 1989, 1992). Each node was planned to have all the necessary social facilities such as schools, shopping areas, recreation, and health centres, so that people would not have to go to other nodes for these services. In view of a more evenly loaded transport system, there was also to be a balance between the number of jobs and the number of people in each node. The detailed land use within these nodes was to be determined at the time of development of each node. By 1991, New Bombay was planned to have 20 such nodes, with a total population of about 2 million and an employment potential of about 800,000 persons (CIDCO 1973:17-25). The sequence in which nodes would be taken up for development would depend on the construction of various transport arteries.

Land use pattern

So far as the proposed land use pattern for New Bombay was concerned, if compared to other cities and towns in India, some differences could be observed.

Table 5.1: Proposed land use pattern in New Bombay (1973) compared to other Indian cities (%)

	Resid	Comm	Public/ semi-pub	Parks/ Playgr.	Roads/ Rlways	Other	Tot.	Industry as % of 1-6
New Bombay	*34.80*	*7.93*	*15.77*	*9.81*	*20.65*	*11.04*	*100*	*16.00*
Gr. Bombay	38.53	9.15	18.40	8.92	12.20	12.80	100	14.92
Delhi	45.46	2.46	18.75	25.10	8.21	NA	100	5.78
Bangalore	56.30	3.20	12.00	12.30	NA	16.20	100	16.40
Bhopal	68.00	2.74	10.10	5.15	NA	14.01	100	11.95
Kanpur	43.70	3.97	17.80	16.02	12.38	6.13	100	13.80
Patna	78.55	2.16	9.26	4.45	NA	5.58	100	6.05
Mangalore	48.73	4.62	11.55	15.80	19.30	NA	100	5.35
Lucknow	52.80	7.20	14.40	14.50	11.10	NA	100	12.50
Aurangabad	35.29	2.36	37.10	11.80	13.45	NA	100	8.67

Source: NBDP 1973. NA=not available. *Others* include service industries and warehousing, university, and State Reserve Police and other Civil Defence Agencies. Note: this table is an exact copy of the one that was included in the NBDP of 1973. It will be noticed that it is not entirely clear how the figures add up and where *industry* fits in. This observation does not, however, obstruct a comparison in land use pattern between different Indian cities.

Land use for residential purposes was to be relatively low, whereas land use for industry and transportation infrastructure was to be relatively high. Since much of the MIDC areas of Taloja and the Thana-Belapur industrial belt were still vacant at the time, it was suggested that these areas would become available 'not for establishing basic industries, but for accommodating industries shifting out of Greater Bombay' (CIDCO 1973:41). The areas already under forest or hills were outlined as regional parks where recreational activities could be developed. An area of 9 km² was being set aside for sewage farming of vegetables and flowers, and it was expected that the sewage of approximately half a million population could be used for this purpose. For fisheries, an additional 9 km² was provided near the Karanja creek.

Broad physical planning

The major job centres were planned at specific locations 'taking into account the existing and future availability of infrastructure, proximity to clients, availability of Jawaharlal Nehru Port (JNPT), natural gas and also the economies associated with agglomeration' (CIDCO 1992:11). Industrial employment was planned in the already existing TBIA and around the new

Nhava Sheva port. Unlike Bombay, residential zones were to be built near, or at least within easy reach of the work place. Therefore, the best location for residential concentrations would have to be central to these three areas, i.e. around the Vaghivali island, in the centre of New Bombay. This area was also believed to be the best location for office activities, not only for socio-economic reasons, but also for environmental considerations, as it was planned to eventually dam the Panvel creek and to create an inland lake over the Vaghivali island. From this central area, four mass transit corridors would be developed in all directions, with two long arms towards Thana (N) and Uran (SW), and with two short arms towards Taloja (central NE) and Panvel (central SE).

Housing

Accommodation was to be made available 'for the entire range of income groups expected in the city', with the intention that 'there should be no *rich* nodes and no *poor* nodes' (CIDCO 1973:17). It was expected by the planners 'that this accommodation of residents from various social and income groups within the same physical area will not only make for a healthier urban environment, but will also ensure a uniform standard of maintenance of social and urban infrastructure services in all neighbourhoods so that no one class of residents is better served than another' (CIDCO 1973:18). Moreover, it was assured that 'every family living in New Bombay shall have a dwelling of its own, however small' (CIDCO 1973:18). Given the average of five members per household for the GBMC and the target population of 2 million, the number of dwelling units required would be about 400,000. The planners realised that the main problem would not be constructing that target number, but rather the household funds available for spending on housing. If household income figures for Greater Bombay in 1971 were to be taken, and given the fact that the average construction cost of *pucca* building (bricks or semi-concrete) was about Rs. 500/m², it would not be possible for 16% of the prospected total population of New Bombay to get anything more than a plot of land of about 30m² per household. For the lowest income group, a further 20% of the population, it would not be possible to provide even this (CIDCO 1973:56).

  Aware of the consequences of a similar situation in Bombay, where a large segment of the population was living in conditions of extreme degradation, CIDCO proposed two important measures to deal with the situation: first, a system of cross subsidies, whereby housing for the upper income groups would be sold with a slight surcharge, which would then be used to subsidise low income housing,[14] and secondly the development of 'sites and services schemes'. Statistical projection had predicted that even after providing such subsidies, 20% of total expected population would still be unable to finance any kind of *pucca* dwelling. For these lowest income households the NTDA would develop plots of approximately 30 m² each, on which households would be permitted to construct huts with the materials of their choice, and in accordance with their financial situation. Since the space between the *sites* would be properly laid out, paved and provided with street lighting, and since they would be *serviced* with water, electricity and sanitary facilities, it was believed that 'despite the cheap construction these developments will be very much better, environmentally, than typical hutment areas in Greater Bombay' (CIDCO 1973:57). However, the main feature of these sites and services schemes was that families were expected and encouraged to improve their initially temporary dwellings into increasingly permanent constructions, for which they could receive the necessary help (provision of construction materials on a retail scale, practical help, etc.). Of utmost importance in this scheme was the stipulation of the planners that 'precautions will be taken to ensure that families which are allotted subsidised plots ... do not sell out to speculators, who would then rent out the same plots at an exorbitant price' (CIDCO 1973:57). To guarantee security of tenure, the plots were allotted on the normal 60 years residential lease.[15]

For the middle and higher income groups the concept of 'row housing' was proposed, with plots being long and narrow, and each family being permitted to build up to the common boundary with its neighbour. The advantages of this kind of housing were considered multiple, the most important being the lower construction cost per square meter and the possibility for families to expand their accommodation according to their needs and finances over time.[16] The case for row housing was even considered to become stronger, as it was expected at the time that over 80% of the future households which would be living in New Bombay would live in units of 25 m² or less (CIDCO 1973:60). Some private space in front or behind such small units was considered to be extremely valuable. However, the NTDA realised that 'as land prices rise and if FSIs are increased, row housing may no longer seem as economical' (CIDCO 1973:60). Notwithstanding the enthusiasm among the planners about row housing, ultimately the public, by their choice and financial capabilities, would decide on which type of housing they would live in, be it row housing, apartment buildings or low rise buildings, which were all to be provided by CIDCO.[17]

In order to bring down the overall cost of housing, CIDCO has been approaching the Housing and Urban Development Corporation (HUDCO) and the Life Insurance Corporation (LIC) for long-term, low interest bearing finance for construction of dwelling units, as well as private banks to invest in housing loans for individual households. Private housing construction was welcomed, although plots of land for residential and commercial use were not to be sold outright, but to be leased for the normal period of 60 years. The NTDA did not expect these private agencies to construct low-income housing, and realised that that role was probably to be taken over completely by itself. In order to curb speculation, the NTDA required that anyone securing a plot of land would start construction within a particular time and would complete it within a reasonable time.

Transportation

The transport system was considered crucial in determining the life style of people. One of the main features of the concept of nodal settlement was to develop nodes along a number of major transport corridors which as we said above, were to start from the central area around the Vaghivali island, running in four different directions. The mass rapid transit system was to be so designed as to be located within a distance of maximum one kilometre for most users. In the initial stage, buses were to function as the only mode of public transport, after some time possibly on reserved tracks (busways). With demand increasing, public buses would become too expensive, and train systems would be introduced in areas where a sufficient demand for such services was available. A normal road system was to be constructed inter-connecting nodes and penetrating such nodes. Since cycling is the cheapest form of transport, CIDCO proposed to construct an experimental cycle track. Pedestrians were to be segregated as much as possible from motorised traffic. In order to persuade offices to move to New Bombay, a rapid transportation service between Bombay and New Bombay was to be provided. As travel times with conventional transport would take at least an hour, an alternative fast water ferry service by hovercraft or hydrofoil was proposed; an idea which at the time was probably more developed for psychological reasons, in order to assure offices that they would not be completely isolated after shifting to New Bombay.

Utilities

The higher the concentration of population within a given area, the lower the cost of providing utilities. The NTDA has always planned its activities on the basis of this generally accepted principle. So far as water supply was concerned, the aim was 'at providing adequate water and a 24 hour supply for all activities, and at making the water bacteriologically pure' (CIDCO 1973:64). The existing water supply system in New Bombay at the time had been provided by the MIDC, which would also undertake any additional water supply, to be tapped from four rivers. Charges for domestic water consumption were to be lower than charges for industries and commercial activities. So far as sewage management was concerned, sewage farming was to take care of the produce of 500,000 people, but expansion schemes were to be implemented when needed. Although garbage collection and garbage disposal were still considered 'a statutory obligation of civic authorities in India', contracting out to agencies was being considered (CIDCO 1973:76). For final disposal it was decided to use the so-called 'sanitary land fill method' (because less expensive). The Maharashtra State Electricity Board (MSEB) was appointed to supply electricity in the new town. New Bombay was already having two major receiving stations, but it was planned to have eight sub-stations of varying capacity, which would make power supply more than sufficient and guaranteed *cut-free*. Post and telegraph facilities were to be provided on the basis of one for every 5,000 to 10,000 population, as was the criterion elsewhere, with a minimum of one per node. Every node was to have a fire brigade station. Telephone lines were to be provided by Bombay Telephones (as in Greater Bombay) in close relation with the NTDA. Typical town amenities such as police stations, banks, market and shopping centres, hotels and restaurants, social centres, public libraries, auditoria and cinema halls, swimming pools, night shelters, crematoria and so on, were to be provided on a ratio basis, but with a certain minimum per node, according to the amenity.[18]

Social services

As the provision of social services such as education, medical care, social welfare or recreation was one of the broad objectives put forward by the state government, it has always been the NTDA's intention to pay as much attention to the social aspects of development as to the physical aspects. The target was 'to ensure that social services in the new city are so developed and managed that every household benefit (*sic*) from them, fully and equitably' (CIDCO 1973:77). The land reservations proposed for social services were expected to be adequate for a long time to come, and financing of such services was so planned that their usage would not be related to the paying capacity of a person. The idea was that the NTDA would provide the broad framework of social servicing, and would actively work together in this field with voluntary agencies and other government bodies.

The school was considered the main hub of social planning, around which all social services in a given neighbourhood were to revolve. The ratio of schools over population was put at the same level as elsewhere in India.[19] Schools were to be so located that children would not need buses to go to school, nor have to cross any highway or major road. Private institutions for education would be welcomed and provided with appropriate facilities, and it was desired that as much education as possible would be covered by such private institutions. The school system would be designed 'to provide a quality of education to all pupils comparable to that currently available only in expensive and exclusive schools in Greater Bombay' (CIDCO 1973:79). The cost of this highly improved school system was expected to be relatively high, but parents would not be required to meet the cost all by themselves. Pre-primary and primary education were proposed to be free, as these were the exclusive and total responsibility of the civic authorities. Secondary schools and post secondary education,[20] like social welfare institutions and hospitals, were possibly to get partial grants from various state government departments to cover salary and overhead costs according to a specific scale. Consequently, if the overall cost for families was to be kept within manageable limits, the NTDA was bound to raise its own resources for education. The planners were thinking of raising these finances through imposition of three kinds of taxation: a payroll tax and sales tax (by the state government), and an establishment tax (by the local authority, i.e. CIDCO). If a sufficient number of institutions of higher learning were to locate in New Bombay, the possibility of developing a separate node for educational and research purposes would be considered. A proposed university was to be located in this node.

Health and medical care were considered another very important social service to be provided by the NTDA, in close relation with potable water, sanitation, environment, recreation, and afforestation. The orientation of public health services was to be shifted from mere prevention to positive action, and had to be so organised as to go out and reach people rather than having people queuing up for attention. Medical care was to be both adequate and of high quality and had to be available to all 'irrespective of the paying capacity of patients'. A three-tier organisation of the health service was proposed, with the Community Health Centre at the local level, the Medical Centre as a kind of small hospital with a polyclinic, and the Hospital for more intensive medical care. The number of hospital beds per 1000 population was put at 4.5.

A third important social service to be provided was a social welfare scheme. As the Government of Maharashtra at the time was planning to expand the social welfare scheme of the state, without there being real scope for expansion of such services in Greater Bombay due to lack of suitable land, it was proposed by the planners that additional institutions for handicapped people, socially exploited women, leprosy-affected persons, juvenile delinquents, or centres for the blind could be conveniently located in New Bombay. Moreover, community centres were to be developed in all nodes and neighbourhoods of New Bombay, offering the services that would be most needed by the community. They were also proposed to be used for income generation activities, especially for low and middle income groups, and were to provide facilities for recreation and social education, reading rooms, or family counselling.

A fourth major social service to be provided was recreation. About 10% of New Bombay's overall land use was reserved for parks and forest, and within the nodes adequate provision for green areas, open spaces and parks was planned.[21] All schools were provided with playgrounds which were to be shared with the local community. Community centres were to be used for indoor recreation. Gymnasia, swimming pools and other sports facilities were to be provided. The Vaghivali lake was planned to be used for water recreation (among other things) and the hilly areas for hiking.

Rehabilitation programme

Integration and rehabilitation programmes for the local people in the villages, and more specifically for those families whose land was to be acquired, were to be an integral part of New Bombay's social services programme. Furthermore, although CIDCO was officially not to undertake any development work in the villages, the Development Plan stated that the villages were to be kept intact and were to be actively developed 'by providing social amenities and otherwise encouraging the process of absorption of the rural population into the new urban setting, to enable them to participate fully and actively in its economic and social life' (CIDCO 1973:103). The rehabilitation programme was (1) to provide at least one member of every family whose land would be acquired with a real opportunity of earning a non-agricultural livelihood at least as remunerative as the family's land afforded; (2) to provide for those among the dispossessed who would not be able to work some secured income from investments out of the compensation payable; and (3) to bring about the assimilation of the rural population in the new urban environment and equip them to take full advantage of the new opportunities created by the development of the new city (CIDCO 1973:103). Every landowner whose land would be acquired was to receive financial compensation, and until actual development work on his land would start, he was allowed to continue his agricultural activities.

With the land, these people would also lose their means of support and social security. To this end, the NTDA proposed a series of rehabilitation measures and schemes for the Project Affected Persons (PAPs): the Corporation was to remain in constant touch with the PAPs by paying regular visits to the villages; technical training was to be offered to the majority of uneducated and unskilled people in specially established Technical Training Centres; a placement service was to be organised; an Employment Guarantee Scheme had already been introduced in several villages to employ PAPs on a daily basis in afforestation schemes or in construction works; an Employment Exchange Programme had been launched by the state government to give preference to PAPs in employment, and an Entrepreneurs' Scheme was introduced to provide jobs to unemployed graduates; loans were to be provided for PAPs who desired to start a small business (a bakery, a grocery, a fruit stall, etc.); PAPs were to be trained in autorickshaw driving, after which loans were to be provided and autorickshaws to be made available; compensation money could be deposited with the NTDA at a high interest; facilities for training in various skills required for employment in industries, in offices, in gardening and so on were to be arranged, and last but not least, special financial measures were to be taken to educate the children of PAPs, and primary schools were to be constructed in all the 95 villages of the project area.

Apart from rehabilitating these PAPs, the villages 'which will be exposed to the great impact of the proposed urban development, which will change their entire life style' (CIDCO 1973:107), were also to be integrated to some extent in the new urban setting, without causing too many changes in the basic village structure and too much disturbance to village life. To this end, dwellings in the villages had to be brought on a par with the urban environment, and several large schemes for improvement of the *gaothans* were to be formulated based on in-depth studies and in close interaction with the village people: a scheme for providing piped water was to be completely financed by CIDCO, with the village panchayat paying the monthly charges; primary health units were to be constructed so as to be available within a reasonable distance of every village; TV sets were to be installed in secondary schools; and film shows were to be arranged periodically.

The implementation and extent of success of these rehabilitation programmes will be discussed in detail in chapter 7.

Financial planning and land disposal policy

'New Bombay', the Development Plan stated, 'may well turn out to be the largest planned city project so far undertaken anywhere' (CIDCO 1973:9). If a project of this scale was to work, the financial planning was obviously of paramount importance. When CIDCO was designated as the NTDA for New Bombay, the Corporation had an authorised capital of Rs. 5 crores and a subscribed capital of Rs. 3.95 crores (Rs. 50 million and Rs. 39.5 million resp.).[22] Although CIDCO was a fully owned limited company of the state government, and could in the initial stage work with government or government-related funds and loans,[23] it was the intention that it should become self-financing and was to attract the major part of its working finances from loans on the private market. The Corporation, thus, was to become a profit-making firm, and had to think and plan as a private company. Finances for the initial years were, therefore, in part drawn from the sale of equity shares.[24]

On 24 January 1972, the state government passed a resolution empowering CIDCO to dispose of land in New Bombay after its development. The sale of developed land was hence to become the soul-saving factor in CIDCO's financial strategy for development of the new town. The overall concept was such that all private land was first to be acquired by the state government, that all acquired private land and publicly owned land would subsequently be transferred to the NTDA, which would develop the land and provide it with all the necessary amenities and infrastructure, and that finally the developed land would be sold off. Apart from public funds and public and private loans, the programme for future expenditure was thus greatly dependent upon the success of the land disposal policy and on the capacity to generate finances on a 'revolving fund' basis.[25]

First development work in Vashi node

Having passed the preliminary planning stage, the NTDA designated an area of 365 ha around Vashi village (comprising the island villages of Vashi and Jui, and the village of Thurbe on the mainland) as the first so-called Early Development Area (EDA).[26] Since Vashi was the first node to be developed, we shall briefly look into the process of initial development of this node, as it shows off fairly well how the development concept was being put to practice on the ground.

Although Vashi's residents were at the time (1972) mainly engaged in maritime activities, during the monsoon season some paddy was being cultivated in the fields adjacent to the villages. The cost for acquisition of these agricultural lands was estimated at about Rs. 4.1 million, but acquisition proposals were received with a lot of hostility and opposition, which seriously delayed the implementation process (see chapter 7). The cost of land in the Vashi area in August 1972 was Rs 6,000/ha for marshy land, Rs. 12,000/ha for firm land in Jui, and Rs. 24,000/ha for firm land in Vashi (CIDCO 1972). About 30% of the land was low-lying and marsh land, and one of the first tasks CIDCO had to take care off was therefore to be in land reclamation.[27] Nevertheless, the choice of Vashi as EDA was fairly logical. It was conveniently located in central New Bombay, close to the already existing National Highway No 4, and at a short distance from Greater Bombay over water. The construction of a bridge connecting the old and the new was most suitable at that location. Furthermore, Vashi was also close to the already existing Thana-Belapur industrial area, which at that time had a very unbalanced development, consisting almost exclusively of industrial units without any urban facilities to support them. Consequently, industrial employees had to commute to and from the workplace, resulting in loss of time and energy.[28] Construction of large housing schemes and provision of basic facilities in Vashi was to solve this problem. It was expected that if residential accommodation was made available at reasonable rates, the demand for housing would easily grow to around 10,000 tenements (about 50,000 people). Housing was therefore considered crucial, and a major potential catalyst for growth and further development in the area. The importance of housing is also reflected by the land use plan that was proposed for the EDA in August 1972. Not less than 63% of the then total Vashi area was reserved for residential purposes (CIDCO 1972:14). Beside for housing, Vashi's setting was also considered ideal for the location of a number of wholesale markets (which were located in very congested areas of South Bombay with not the slightest possibility of expansion), as well as for cold storage and warehousing. Consequently, another 28% of land was being reserved for this purpose.

As mentioned earlier, in terms of physical planning, every node was to be sub-divided into sectors, and for Vashi the plan was that each sector was to have a population of 6,000 to 7,000, consisting of a well-integrated socio-economic mix of various types of residents. Eventually, the Vashi township was to contain about 100,000 people. A primary school, shops and recreation facilities were considered as an absolute minimum to attract some of the industrial employees still residing in Greater Bombay and Thana, and to bring about some urban growth.

The NTDA was having three major capital expenditure posts in Vashi, viz. acquisition and reclamation, physical infrastructure, and social infrastructure. The total cost of physical infrastructure (roads, storm water disposal systems, water and sewerage system, street lighting, landscaping, etc.) was estimated at Rs. 84 million, and that of social infrastructure (education, markets, hospitals, community centres, cultural and recreational facilities, religious facilities, and so on) at Rs. 19.9 million.[29] In terms of expenditure distribution this was about 65% and 15% respectively of total expenditure, with the remaining 20% to be spend on acquisition of land. Since CIDCO had to become financially self-sufficient, these costs had in some way or another to be recovered. The major part of this was obviously to come from the disposal of developed plots of land.

The pricing policy for land thus had to cover the cost of development and had to help the project to reach self-sufficiency as soon as possible. On the other hand, the pricing policy was not to take prices beyond the reach of an average man in New Bombay. After exclusion of the land that was needed for the physical and social infrastructure, only 187 ha out of the total of 365 ha would be available for disposal. The entire cost of the Vashi development project divided by the net disposable area, resulted in a land price of Rs. 64/m², which was not much less than prices in many areas of the Bombay Suburbs at that time. Consequently, since residents in Greater Bombay or new incoming migrants were not going to be attracted by such small differences in land prices, the so-called 'reserve price' (RP) had to be pushed lower. Land lease premiums for commercial and industrial use were to be 15% (industrial) to 60% (commercial) higher than for residential use.

Capital spent on traffic and on utilities was to be recovered separately. The expenses on road transport and electricity would be reflected in the rates for individual use of these facilities, but use of the same system would make water far too expensive, and expenses on water were therefore to be recovered through lease premia on land, service charges in lieu of local taxation and through water rates (CIDCO 1972:78-86). In exchange, an excellent service was to be expected, as the supply of water was planned to be 'available round the clock and without the need for installing pumps ... except in the case of high rise structures' (CIDCO 1972:43). A per capita domestic supply of about 225 litres/day would ultimately be achieved, and CIDCO's intention was 'to over provide rather than to fall short' (ibid.). All in all, compared to Bombay, where several taxes were being levied, 'taxation' in Vashi was in the early 1970s about 40% of that in Bombay proper (CIDCO 1972:78-86).

Implementation of the planning policies in the EDA of Vashi during the 1970s were anything but successful. In the initial years, a lot of time was spent, and sometimes wasted: on a number of studies which had to be conducted before moving on with the implementation,[30] on the acquisition of the land due to heavy opposition from the local people (see chapter 7); and on problems in the CIDCO head-office, as well as in the field.[31] Institutionally, there was a serious conflict between the NTDA and the state government over the payment of compensation money, which heavily delayed the planning and development work.[32] In the field, on the other hand, there were not only the problems with the PAPs concerning acquisition arrangements, service charges, etc.,[33] but also increasing popular criticism on the NTDA, in part because of its internal strife, and the Corporation was pressed hard to actually start doing something, even in the light of acquisition difficulties. It was therefore ultimately decided to evade the main problem, and to start with a first large housing scheme on reclaimed land in Vashi. Unfortunately, on the one hand the proximity to Bombay already created the impression of a suburban satellite town rather than that of a self-sufficient town, and on the other hand the cost of reclamation was prohibitively high and created a liquidity problem. An attempt to convince the industrial workers of the TBIA to take residence in the area, which would have offered a solution, failed, and CIDCO increasingly became a financial drag on the state government and was finally put completely under government control again in 1976.

Since the monetary pressures prevailed, the Vashi accommodation was ultimately thrown open for sale on the private market. It was a major deviation of what had been planned, and a first step in a process of increasing use of market principles within the NTDA. The situation improved considerably from 1979, the year in which the Draft Development Plan was finally approved. A lot of money had been brought into the region due to a number of large contracts with state and central government agencies and companies,[34] and hence more finances became available for a decent compensation for the affected people. The decision by the state government to put the compensation money at Rs. 15,000 per acre, caused a sudden spurt in land prices. It was the start of a spiralling inflation in New Bombay's land market, reinforced by CIDCO's policy of disposing of land to the highest bidder. Consequently, New Bombay increasingly started to attract developers and investors who had previously been focusing exclusively on Greater Bombay. It was only then, by the early 1980s, that urban growth and development in New Bombay could actually start.

Conclusion

In the early 1970s, at the time of notification of the New Bombay area, the 25,000 odd households living in the 95 villages in the area were mainly enganged in agriculture, fishing and salt pan labor. Over half of all the land was privately owned by a large number of families and about a third was publicly owned. The older municipal towns of Panvel and Uran in southern New Bombay were the only locations in the area already having certain urban characteristics. The industrial belt between Thana in the north and Belapur in central New Bombay was slowly developing. Most of the industries in this belt were chemical, capital-intensive industries. Essential urban facilities however were lacking, and the large majority of industrial workers were outsiders who commuted to and from New Bombay. Partly as a result of this (limited) chemical production in New Bombay, and partly due to the industries of the Bombay Suburbs, the area was already experiencing a growing problem of pollution.

In March 1971 the City and Industrial Development Corporation of Maharashtra Ltd. (CIDCO) became the New Town Development Authority, responsible for the planning, implementation and maintenance of New Bombay (and until a separate municipal body would be created also for the administration of the new town). Mid-1973 the NTDA published its New Bombay Draft Development Plan, which had to be on a par with the broad development objectives of the state government. The plan was a detailed blueprint of what New Bombay was to become and how, and put suggestions, guidelines and proposals forward on the new city's overall development concept, its land use pattern, its broad physical and social planning, and the rehabilitation of the local village population. Financially, development of the area was to be such that, apart from initial seed capital, urban growth could be induced on a basis of financial self-sufficiency. Land was to be treated as the main resource. It was to be acquired, to be developed with all the necessary infrastructure and amenities, and to be sold off in order to generate the necessary funds and to make further investment in urban development possible. Or, as L.C. Gupta, a former Managing Director of CIDCO has put it, while developing New Bombay, CIDCO was 'required to recoup all the costs of development and acquisition so that the development of a city takes place on a 'No profit no loss basis' (Gupta 1985-b:3).

Thus, although the NTDA was created as a fully owned company of the state government, it had to function mainly along private business lines. After the initial failure in the first five or six years, when the entire New Bombay idea did not seem to create the minimum of enthusiasm that was hoped for, not among potential investors and developers, nor among potential inhabitants, the shift away from public sector working mechanisms to private market principles was necessarily increased with the sale of the already built housing stock on the private market. This move was absolutely necessary to avoid a financial disaster and to prevent a complete collapse of the entire project. This, at least, was the official statement.

Notes

[1] The *new town concept* is one of the strategies that has, since the 1960s, been developed and implemented in most Third World countries, in an attempt to create a more balanced urbanisation pattern.

As has been argued in chapter 2 of the thesis, the predominant view on overall development in the 1950s and 1960s was based on the belief that cities played a generative role, bringing development to backward rural areas. Therefore, establishing new urban entities at strategic locations was considered the right approach to be followed. Depending on their size and location and their proposed specific function, these new urban places were called 'growth poles', 'growth centres', or 'new towns'. Notwithstanding the fact that the growth pole concept was very broad and could cover a wide range of development strategies, in practice very little was actually done. Few of these new towns became fairly large and/or important urban areas, in part, as Gilbert and Gugler argue, because growth pole strategies did not sufficiently deal with the problem of urban economies of scale, and existing larger settlements probably offered more efficient use of economic resources (Gilbert and Gugler 1984:177).

Although a large number of these new towns did not significantly take off and did not develop into something more than a middle-size town of second or third rank, one category of new towns that had reasonable success was the category of planned new capital cities. The first Third World countries which implemented the idea to create a new capital city from scratch, far away from the then existing capital city, were Turkey in 1923 and Brazil in 1956, the latter which gave Oscar Niemeyer *carte blanche* to develop its new capital Brasilia. Many similar mega-projects have been planned since, and to varying degrees implemented. Some of these no-one has ever heard of, and they are merely capital cities in a formal sense. Others, such as Brasilia or Islamabad, have fully developed into important cities. The experience with these new capital cities shows that if such new towns have to become more than just a symbol of personal or regional prestige, they must be planned as a metropolis with a fully developed economic life, and with a gradual but synchronous development at different levels (infrastructurally, administratively, economically, socially and culturally).

Over the years, a debate has been going on between those in favour of the new town concept, and those like Renaud, who considers new towns as absolutely useless for Third World regions, and a complete waste of scarce financial resources.

---

[2] The list of the 95 villages included in the New Bombay area can be found in the government notification of 20 March 1971 (Urban Development, Public Health and Housing Department). The state of Maharashtra is organised upon several administrative layers. The state is divided into six 'Divisions', each having a specific number of 'Districts' or 'zillahs' (30 in total in Maharahtra), with every District comprising a number of 'tehsils', which in turn group a number of 'villages'. Of the 95 villages that were designated, 29 belonged to the Thana Tehsil of Thana District, 38 villages to the Panvel Tehsil of Raigad District, and the remaining 28 villages to the Uran Tehsil of Raigad District. Raigad District was formerly known as Kolaba District.

[3] A similar concept was later adopted in Shanghai, with the development of the new town Pudong.

[4] For details on legal provisions and organisational regulations, see *City and Industrial Development Corporation of Maharashtra Limited, Memorandum of Association and Articles of Association* (CIDCO, May 1970).

[5] The setting up of New Town Development Authorities was envisaged in the amended Town Planning Act 1966 for implementing proposals of setting up new towns within a region. The concept was modelled after the British New Town Development Corporation.

[6] J.B D'Souza (IAS) was appointed the first Managing Director. He had a planning team of seven people in hand, in which Shirish Patel (one of the authors of the MARG Piece) became in charge of planning the new city. Charles Correa, another of the three authors of the MARG Piece became the Chief Architect. The entire planning team was composed of private consultants, who were privately and (apart from Patel) part-time employed.

[7] In later years, CIDCO was also assigned other major urban development projects in the state of Maharashtra and outside. At present, the Corporation is active in developing new towns in New Aurangabad (since July 1973) and Waluj (early '92) in central Maharashtra, in New Nashik (since March 1977) and Vasai Virar (since May 1990) north of Bombay, in New Nagpur in the eastern corner of the state (since 1992/93), in New Nanded in the south-eastern part about 300 kms. from Hyderabad (since March 1977), in New Solapur and New Kolhapur in South and Southwest Maharashtra, and to a lesser extent in Oros along the southwestern coastal line (since March 1989). Outside of Maharashtra, CIDCO was recently assigned part of the redevelopment project in Cochin (Kerala).

[8] The first elections for the NMMC took place in January 1995.

[9] From a discussion with A.B. Sapre (5 Feb 1995).

[10] Since 1955, Panvel and Uran were two out of nine District Municipalities governed under Bombay District Municipal Act 1901. The state government has the power to declare by notification any local area to be a Municipal District, and also to alter the limits of any existing Municipal District.

[11] In all official planning documents, the villages are called *gaothans* (Marathi), having the same broad characteristics as other typical Indian villages.

[12] In the initial stage, the two main reclamation schemes were 250 ha of land in Vashi node, and 100 ha of land in Belapur node. Both areas were to be reclaimed by the *Dutch Method* of building dykes, using pumps and allowing the water to flow at low tide and block it from coming in at high tide. This method was considered more appropriate than the conventional method of raising the level of lands with earth filling methods, since both areas were under the tidal range and would otherwise have meant about 1.5 and 2 meters of filling (from an interview with Shirish Patel, first Chief Planner of CIDCO) (23 Feb 1996).

[13] The concept was to some extent modelled after Harris and Ullman's multiple nuclei model of urban land use pattern (which assumes that most cities represent an amalgam of multiple nuclei which have functional specialisations), and after the British decentralizing settlement pattern and the construction of new towns such as Milton Keynes.

[14] The lowest income groups would receive the highest percentage subsidy, the highest income groups would pay the highest percentage surcharge.

[15] The implementation of these 'sites and services schemes', and the extent to which they were successful, will be discussed in chapter 8.

[16] Row housing is much cheaper for several reasons : some walls are shared with neighbours, there is no wasted space in the form of publicly shared entrance foyers, staircases and landings, row housing consumes less cement and steel per unit, they can be built more speedily, and in contrast to large apartment buildings they are open to small contractors also, who usually work cheaper.

[17] The precise amount of housing was in 1973 not yet clear.

[18] For example, one police *chowki* (post) per 15,000 population and at least one police station per node, one bank per 10,000 population spread in proportion of the expected population in each area, one shop per 66 persons in residential areas, one cinema seat per 50 persons, etc.

[19] For a population of 2 million this was to be 100 units for pre-primary education, 360 units for primary school, 257 for secondary school, and 46 for colleges.

[20] Post secondary education comprises colleges of arts, schience and liberal education, technical institutes, professional and vocational education, institutes of higher learning and research, university.

[21] This 10% proposal for greenery was rather low compared to some other major cities (see table 5.1).

[22] The financial operations of CIDCO started in March 1970, with a paid-up capital of Rs. 10 lakhs (1 million) fully subscribed by the holding company (SICOM) and a loan of Rs 270 lakhs (20.7 million) sanctioned by SICOM.

[23] It was, for example, the state government who financed the cost of acquisition of land, but CIDCO was expected, in due course, to repay this investment with interest.

---

[24] The number of equity shares (of Rs. 100 each) that was issued went up from 10,000 in the first financial year, to 300,000 in the second year.

[25] Generating finances on a revolving fund basis had in India for the first time been done by the Delhi Development Authority.

[26] By May 1992, the total gross area of Vashi would have increased by exactly three times.

[27] The Board of Directors sanctioned the proposal to start with reclamation of land, in order to connect the two islands of Vashi and Jui, on 31 March 1971.

[28] It was estimated that transportation from Thana station to work was costing companies about Rs. 50 per head per month, whereas industrial development in the area was constantly increasing, with many industries having blueprints for expansion (CIDCO 1972:8).

[29] There was (and is) no clear-cut definition of which items exactly fall under *physical infrastructure* and which under *social infrastructure*. Fire brigade requirements, for instance, are sometimes considered as physical, sometimes as social, which makes cost comparison a rather dubious activity. Social facilities are usually kept apart mainly because their financing often comes from private sources and not from the tax payer.

[30] The Soil Investigation Study; the Water Resource Study; the study on the Economic Profile of Greater Bombay; the Industrial Location Study of Maharashtra; the Air Pollution and Environmental Study in New Bombay.

[31] According to one CIDCO-officer, the inertia during the first five years was actually more due to politics. According to this source, there were simply a lot more important things to be done (the war with Pakistan, the winning of general and state elections, etc.). Especially in times of elections nothing would be done, because politicians needed the votes from the PAPs as well as from the landowners.

[32] Facing the opposition from the landowners against the acquisition of their land, the state government proposed to pay more compensation money. D'Souza however, the first Managing Director (MD) of CIDCO (March 1970-Aug 1974) suspected that part of that money would be used in a corruptive way, and he prefered to pay the then market price to the landowners instead. Consequently, there were attempts by the state government to remove D'Souza as MD, and P.C. Nayk was installed as a second MD (named MD-II, to put them both on the same level). Naturally, this unusual construction with two equal MD's fighting each other, not only led to internal inertia, but also trickled down to the lower levels, where work suffered and morale sank.

[33] The New Bombay area was at that time a stronghold of the Peasants and Workers' Party (PWP).

[34] The deal with the ONGC attracted most of the finances during this period.

6 Growth and Development of New Bombay

Introduction

The New Bombay project was launched as a major instrument to relieve Bombay Island from the increasing pressure of people and economic activities and from its parallel processes of congestion, infrastructural problems and urban 'inadequacies'. In the previous chapter we have discussed the major objectives of the project put forward by the state government and, given these objectives, the proposed principal policies, strategies and activities of the NTDA.

In the developed world as well as in the Third World, several plans with a similar aim and with a similar broad planning concept have been put forward in an attempt to relieve congested cities from the pressures of rapid growth and expansion. Very few such projects, especially in the developing world, can be called successful in terms of growth and overall development. Most of such 'new towns' or 'urban growth poles' (Dodoma in Tanzania, Yamoussoukro in Ivory Coast, Viedma/Carmen in Argentina, Beida in Lybia, Abuja in Nigeria,... and in India for example Chandigarh and a whole list of mainly industrial towns) are playing a limited and insignificant role in their national or regional urban development context, or are total failures and are dying a silent death. In many instances they have, therefore, been a waste of scarce financial resources.

In this chapter we shall see what the situation is in New Bombay. On the basis of statistical material we will get a clear picture of the extent and evolution of growth and development in this new town after 25 years of project implementation. Most of the figures are secondary data in the sense that although most of them have been collected, interpreted and tabulated by myself, they were already available in some form or another in government or semi-government offices like CIDCO, the BMRDA, the JNPT, the Railways, etc. Some figures have been taken from official planning documents and from regional or local newspapers. All the data and tables will mention the primary source. It will be noticed that some tables are incomplete and that some data are not very consistent. This is mainly because the NTDA and other government bodies do not gather certain data on a regular basis, or have only recently started collecting such data.

Fields in which data have been gathered are a) population; b) the economy; c) land prices and property rates; d) government activity; e) transportation; and f) social provisions. All the data were collected to serve as indicators of the extent of growth and development in the area. In this chapter we will only discuss the most relevant indicators.

Indicators of growth and development

Population

As mentioned earlier, both the old municipalities of Panvel and Uran and the villages were not included in the planning and development picture, and were already having a significant population before the creation of the new town project.

Table 6.1: Population growth New Bombay 1961-1995

	Rural	Urban (a)	Total
1961	82,312	28,359	110,671
1971	116,789	39,218	156,007
1981	183,593	66,938	250,531
1991	243,901	291,308	535,209
1995 (b)	373,001	326,526	699,527

Figures from CIDCO departments. Notes: (a) 1961 and 1971 *urban* figures are exclusively the population of the municipalities of Panvel and Uran; 1981 and 1991 figures include population of Panvel and Uran, but 1995 figure excludes this segment of the population; (b) estimate; population of Panvel and Uran municipalities (around 60,000) now included in *rural* figure; *urban* figure now called *nodal*.

Because of inclusion and exclusion of the old municipalities in different figures, it is only possible to compare the total population figures over time. The average growth rates per decade for the period 1961-1991 were 4.1%, 6.1% and 11.4% per annum respectively. The estimated average growth rate for 1991-1995 is around 7.7% per annum.

Figure 6.1: Total population of New Bombay, 1961-1995

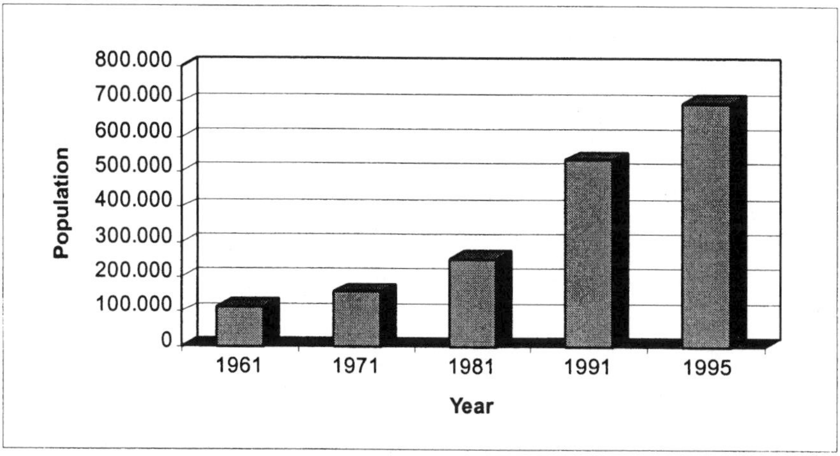

Based on figures table 6.1. Note: 1991-1995 only four-year period

The geographical distribution of the population over the different nodes is a reflection of the level of development in each node. The planning set-up was such, that development in a second node would only start when development in a first node would be nearly completed. Vashi, the first node that was taken up for development, is therefore at present also the most developed node. Currently, eight nodes have been completed or are near completion, all of them in the northern half of the New Bombay area (figure 6.2), in another six nodes work has started at different stages and is in progress, and in the six remaining nodes (most of them in the southern half of the New Bombay area) work is still largely to begin. All people presently living in New Bombay are (apart from the municipalities and village population) spread out over the eight completed or almost completed nodes of northern New Bombay.

Figure 6.2: Eight completed nodes of New Bombay, 1997

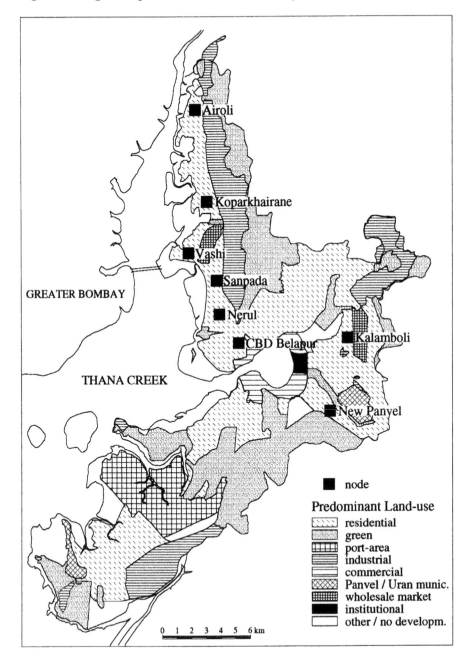

Table 6.2: Population growth per node 1987-1995

	1987	1991	1995
Vashi	72,000	83,178	107,919
Nerul	16,000	36,768	57,597
Airoli	7,000	29,661	40,681
CBD Belapur	18,000	24,302	28,919
New Panvel	10,000	16,864	26,101
Kalamboli	6,000	15,096	29,019
Koparkhairane	---	3,489	30,168
Sanpada	---	381	6,122
Total	129,000	209,739	326,526

Figures: from CIDCO offices. Note: figures 1987 rounded and figures 1995 estimates

Figure 6.3: Population growth 1987-1995 per node

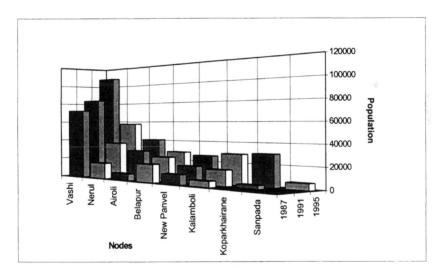

The economy

Industrial and commercial growth and employment General figures on economic growth in New Bombay are not available for the new town separately. The only figures that are available and do give a reasonable indication of overall economic development are those on large-scale industrial investment and projected employment at the District level from the Directorate of Industries (Government of Maharashtra).[1] As the New Bombay area falls under the two Districts of Thana and Raigad, the figures for these two Districts give an indication of the industrial investment pattern in New Bombay as compared to those in other Districts.

Table 6.3: District-wise investment and job creation in large-scale industries in the state of Maharashtra, July 1991 - July 1995 (a)

Divisions Districts	N	Undec.	Investment (Rs. crores)	%	No. of jobs	%
Amravati						
Akola	12		268.72		1,240	
Amravati	6		130.13		3,512	
Buldhana	4		119.46		510	
Yavatmal	15		2,775.59		3,092	
Total	**37**	**1**	**3,293.45**	**3.00**	**8,354**	**3.88**
Aurangabad						
Aurangabad	89		2,417.61		14,601	
Beed	3		13.86		288	
Jalna	12		113.83		1,506	
Latur	8		137.72		487	
Nanded	8		762.90		555	
Osmanabad	3		36.67		100	
Parbhani	2		14.94		173	
Total	**125**	**14**	**3,497.53**	**3.19**	**17,710**	**8.23**
Konkan						
Bombay	143	34	5,114.60		10,835	
Raigad	*230*	*35*	*19,612.72*		*25,037*	*10 %*
Ratnagiri	37	3	17,849.92		10,839	
Sindhudurg	11	0	6,200.98		3,164	
Thana	*372*	*34*	*17,209.71*		*26,999*	*10 %*
Total	**793**	**106**	**65,987.93**	**60.16**	**76,874**	**35.75**

Nagpur						
Bhandara	13		4,462.00		945	
Chandrapur	21		3,507.10		2,305	
Gadchiroli	3		216.18		532	
Nagpur	89		9,597.15		16,039	
Wardha	11		749.10		1,850	
Total	**137**	**4**	**18,531.53**	**16.89**	**21,671**	**10.08**
Nashik						
Ahmednagar	44		1,043.93		11,159	
Dhule	11		216.30		3,036	
Jalgaon	14		307.81		2,741	
Nashik	96		2,347.65		11,736	
Total	**165**	**7**	**3,915.69**	**3.57**	**28,672**	**13.37**
Pune						
Kolhapur	90		1,298.12		12,638	
Pune	219		7,945.20		35,891	
Sangli	14		991.37		4,194	
Satara	13		3,252.20		3,353	
Solapur	33		979.60		5,623	
Total	**369**	**33**	**14,466.49**	**13.19**	**61,699**	**28.69**
Grand total State	**1626**	**165**	**109,692.62**	**100**	**214,980**	**100**

Source: Directorate of Industries, Government of Maharashtra. Notes: names in bold-italic are State Divisions, others are Districts; (a) includes units already producing, under construction and under implementation; note that 165 companies have not yet (1995) decided on the amount of investment, and that the total investment in the state will therefore be significantly higher.

Figure 6.4: Investment and job creation in the six Divisions of Maharashtra (July 1991–July 1995)

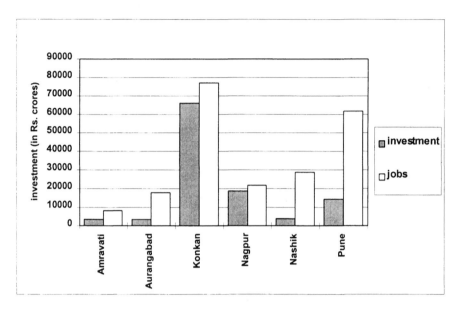

Figure 6.5: Investment and job creation in Konkan Division (1991-1995)

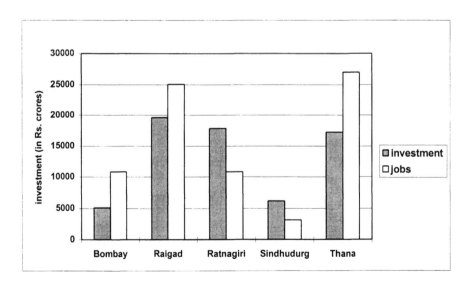

As can be observed from the table (6.3) and graphs (6.4 and 6.5), the amount of large-scale industrial investment in the two Districts of Thana and Raigad was between 1991 and 1995 the highest in the entire state.[2] Significantly, investment in each of these two Districts was higher than total investment in any State Division.[3] What is more, the proposed number of new jobs created in both the Districts of Thana and Raigad between 1991 and 1995 were (apart from Pune District) also the highest in the state, and higher than the total number of jobs created in three other State Divisions! The two Districts of Thana and Raigad have thus clearly been the major focus of large-scale industrial investment in Maharashtra State. Although no strict conclusions can be drawn as to where exactly within the two Districts that money has been invested, it may be assumed that a significant part of it has been flowing into the New Bombay area; an assumption which is reinforced by numerous other data on economic development and employment in New Bombay.

As we have seen earlier, at the time of notification of the New Bombay project, employment in the area was restricted to industrial employment in the TBIA between Thana in the north and Belapur 25 kilometres further south. Since those days, industrial employment, and with it employment in a whole range of commercial activities, has rapidly increased, especially since the early 1990s. As table 6.4 shows, as late as 1987, the total number of workers in New Bombay was still less than 39,000. In the seven following years, however, that number increased by almost 4 times to over 148,000.[4]

Table 6.4: Industrial and commercial employment in New Bombay, 1971-1994

	1971	1987	1990	1994
commercial activities (a)	NAppl.	NAv.	NAv.	44,304 (b)
industry (c)	15,732	NAv.	55,286	103,768 (d)
Total	15,732	38,705 (e)		148,072 (f)

Figures: from CIDCO departments, Bhattacharya (1971) and Survey of Industries (1990). Notes: NAppl. = not applicable, NAv. = not available; (a) commercial activities include small private shops and social activities; (b) does not include CIDCO personnel (about 2000); (c) industry in TBIA, Taloja and port-related industries; (d) employment in TBIA alone 81,296; (e) from CIDCO (1988); (f) excludes the industrial area of the Uran municipality and village employment (15,000-20,000 workers).

**Table 6.5: Estimated total employment per sector and per node
in New Bombay, 1995**

	Vashi	Nerul	CBD	Kalam	New Panvel	KK	Airoli	Khg	Total (000)
Shops / Offices	14,889	2,412	8,391 (*)	1,047	1,362	199	1,218		29.5
Whole-sale	5,937			1,779					7.7
Mafco	3,229								3.2
Truck terminal	328			165					0.5
School/ College	1,221	741	295	202	380	223	497	114	3.7
Hospit/ Nursery	822	64	73		22	25	21		1.0
Social Facils.	301	77	74	12	42		41		0.5
TBIA									81.3
Taloja IA									19.3
Jawah. IA									1.7
Panvel IA									1.4
Total	26,727	3,294	8,833 (*)	3,205	1,806	447	1,777	114	**149. 971**

Source: CIDCO departments. Notes: IA = industrial area; (*) CIDCO employees included

The most recent set of figures (table 6.5) gives the overall employment per sector and per node in 1995 for the entire area.[5] The first seven categories are commercial activities, the last four industrial activities. If the 'commercial sector' is limited to shops and offices, and to the three most important commercial nodes Vashi, Nerul and CBD Belapur, the number of employees per period (from the 1970s to 1994) clearly shows the rapid acceleration in commercial activity since the latter half of the 1980s (figure 6.6).

Figure 6.6: Number of employees in shops and offices (cumulative) per node, 1970-1994

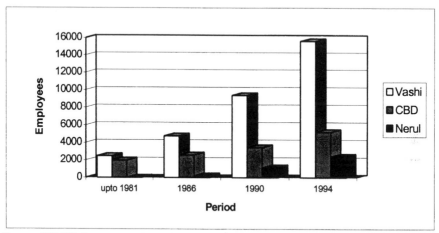

Source: CIDCO departments. Note: 2000 CIDCO employees not included in CBD figures

Although the new CBD in and around Belapur node was planned as the major commercial (and administrative) centre of New Bombay, up to February 1996 commercial employment in the CBD was very limited. Total employment in 1994-95 was between 8,391 and 8,833 (CIDCO employees included), which makes it the node with the highest employment after Vashi, but almost all the functions in the CBD are government-related. As table 6.4 shows, the real commercial centre of New Bombay is not located in this new CBD but in Vashi node.

Wholesale market activity Table 6.4 gives a figure of 7,716 people employed in New Bombay's 'wholesale sector'. In September 1978, the BMRDA accepted a proposal to shift the wholesale agricultural produce markets (APM) from their traditional location in the Masjid Bunder area of South Bombay to a planned and spacious new location at Turbhe, near Vashi in New Bombay.[6] Likewise, the proposal to shift the wholesale iron and steel markets from their traditional location at Carnac Bunder and in the Darukhana area of South Bombay to a new location at Kalamboli in New Bombay was also accepted. This major move of a large economic sector away from Bombay was to have a double purpose. It was to be a major contribution to the decongestion of Bombay Island on the one hand,

and was supposed to be an important catalyst of economic development and activity in the New Bombay area on the other.

In March 1981, the onion and potato market, part of the APM-phase-I, was the first market to be shifted out of South Bombay. Although a lot of opposition was met, and many problems have arisen with shifting the other agricultural markets in the years thereafter, since 1991 things seem to be moving again and more and more markets are shifting their activities to Turbhe. The total trade turnover of all the fourteen agricultural wholesale markets (onions and potatos, grains, pulses, sugar, spices, etc.) that are to be shifted, exceeds Rs 3,000 crores (30 billion) per year. The multiplier effect of shifting this huge economic sector into the New Bombay region is evidently very significant, and activities such as transport, trucking, cold storage, and finance are being attracted and expanded. Construction of a goods railway line between Kalwa in the north and the warehousing complex at Turbhe has been completed and is ready for operation. At present, activity in the agricultural warehousing area is gradually but steadily increasing, all of the trading units have been booked, and according to figures of 1994 nearly 6,000 people are employed in the new APM, with many more in the associated sectors. According to CIDCO's Chief Economist this figure of 6,000 must now be at least 30% higher.

Shifting of the iron and steel markets from South Bombay to Kalamboli in New Bombay has met even more opposition, and although the new market was formally opened in October 1989 and markets have started to shift, warehousing activities in Kalamboli are still in its infancy. The official employment figure here was in 1994 less than 1,800. It is obviously difficult for a sector in which activities are so closely linked to be shifted in stages, but ones an important part is shifted, the others will follow soon. There is no doubt, therefore, that activities and employment in these two wholesale locations of New Bombay will exponentially increase in the years to come, and the wholesale sector is bound to become one of the major economic growth poles in the area.

Port activity The plan to build a new and modern port opposite the old and congested colonial port of the Bombay Port Trust (BPT) had already been proposed long before there was any mention of a New Bombay project. For many years, the plan remained just a plan. Finally, in May 1989 this enormous project was commissioned. The new Jawaharlal Nehru Port (JNPT) has a maritime area of about 52 km² and is said to be the most technologically advanced port of India and one of the most modern ports in the whole of Asia. In the short period since 1989, the new port has developed into an important nodal point of exchange of cargo from sea to surface transport and vice-versa. Total cargo handled at JNPT has increased almost tenfold within a six-year period, from nearly 0.7 million tons in 1989-90 to almost 7 million tons in 1995-96.

Table 6.6: Total traffic at JNPT Port, 1989-1995 (in thousand tons)

Commodity	1989-90	1991-92	1992-93	1993-94	1994-95	1995-96 (*)
Total bulk cargo	290.7	1,443.2	1,279.2	1,280.4	2,034.8	1,800
Total no. of vehicles	Nil	24.6	16.7	23.0	30.1	30
Total containers	406.5	1,313.9	1,711.7	2,076.9	2,928.9	3,800
Total cargo handled	697.3	2,794.7	3,007.4	3,388.2	5,008.0	6,800

Source: M.G. Agire, manager personnel and individual relations JNPT. (*) projection by D.N. Maurya, technical adviser to chairman, JNPT

Overall growth in port traffic during 1994-95 was 47.18%, with an increase of 58.9% in bulk cargo, and 41.0% in container traffic. Exports played a major role in this growth figure, with a 40% increase in the export of vehicles. The JNPT is projected to handle traffic volumes in excess of 11 million tons by the turn of the century, and a large expansion scheme is already in progress.[7] Financially, the JNPT's net surplus increased from Rs. 5.56 crores in 1990-91 to Rs. 62.19 crores in 1994-95 (Rs. 55.6 million and Rs. 621.9 million respectively).[8]

Construction activity Another major economic indicator that gives an idea of the extent of growth and development in New Bombay is residential and commercial construction by both the public (CIDCO) and the private sector, the latter which is obviously the more significant growth indicator. Since the very beginning of its activities, the NTDA has been selling land based on the 'reserve price' (RP).[9] The first Project Report for Vashi (1972-73) put the RP at Rs. 35 per m², by 1983-84 it had gone up to Rs. 252 per m², but went down again afterwards.

A potential buyer of a CIDCO house or land must pay a price based on a percentage of the RP, depending on which income group he belongs to. CIDCO has always considered four income groups: Economic Weaker Sections (EWS), Low Income Groups (LIG), Middle Income Groups

(MIG) and High Income Groups (HIG). Housing constructed by CIDCO is being sold to the EWS at 25% of the RP, to the LIG at 50 to 100% of the RP, to the MIG at 125% of the RP, and to the HIG at 175 to 250% of the RP. Since the mid-1970s, some 95,000 houses have been built by CIDCO, of which about 52% were to cater for the EWS and/or LIG, 27% for the MIG, and 21% for the HIG (Bhattacharya 1995). The sizes of CIDCO housing range from less than 20m² for EWS and LIG, over 21 to 50m² for MIG, to 51m² or more for HIG.

Table 6.7: Housing construction, CIDCO and private (cumulative), 1975-1995

Year	Housing construction		
	CIDCO	Private	Total
1975	1,238		1,238 (a)
1980	3,971 (f)		3,971 (b)
1983	10,428 (f)		10,428 (c)
1987	NAv.	NAv.	51,081 (d)
1991	65,663	19,161	84,824
1995	95,000	40,000	135,000 (e)

Source: 1975-1991 figures CIDCO departments; 1995 figures Bhattacharya (1995). Notes: (a) first housing construction exclusively by CIDCO in Vashi, sector 1; (b) Vashi sectors 1 (CIDCO) to 8 (private); (c) Vashi (both CIDCO and private); (d) construction in six nodes (both CIDCO and private); (e) estimates, and part of it under construction; (f) separate figures for public and private housing not available, but in those years very little private building construction in operation, and the total figure thus refers almost exclusively to CIDCO housing.

Figure 6.7: Housing construction (CIDCO and private), 1975-1995

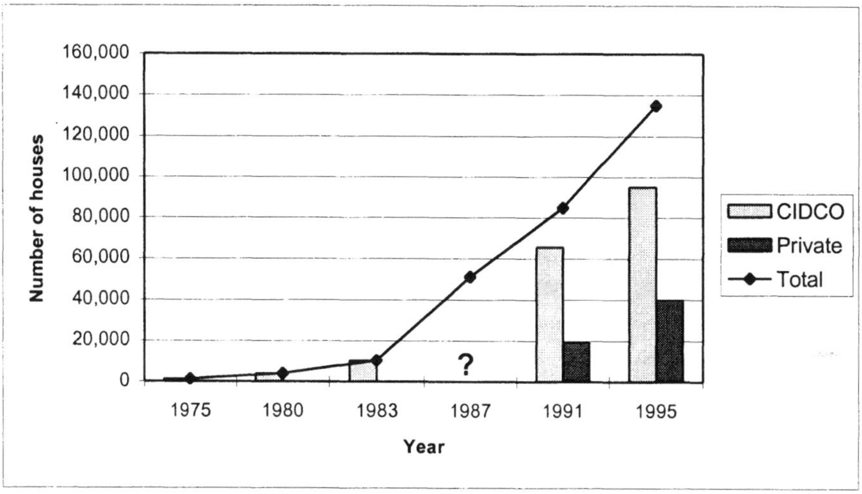

Based on figures table 6.7. Note: no separate figures found for 1987

So far as commercial construction is concerned, the NTDA has sold commercial premises according to whether the building was to function for 'low order shopping' (small shops of less than 10m², in market places for example), 'middle order shopping' (10 to 20m²), or 'high order shopping' (more than 20m²). Shops of less then 10m² are being sold at RP or rented out, shops from 10 to 20 m² are being sold at 400% of RP (fixed price) or by tender, and shops bigger than 20m² are being sold by tender at minimum 600% of RP. Table 6.8 shows the number of private offices and shops that has been constructed since 1991.

Table 6.8: Construction of private offices and shops since 1991 (a)

	Private offices	Shops (b)	
		Private	Cidco
1991	4632	5089	
1992	5213	5471	
1993	5625	5982	5495
1994	6467	9239	

Source: CIDCO statistical department. Note: (a) offices and shops both constructed and under construction; no figures available before 1991; (b) shops include service shops, ice-cold storage, show rooms, construction materials, etc.

Land prices and property rates

Evolution   The evolution in land prices and property rates is, beside population figures, probably the most significant indicator of growth and development in a given area. In New Bombay, the NTDA has been selling land for residential purposes either at 250% of RP (in developing nodes) to 400% of RP (in developed nodes) in case of society plots, or at a bottom price of 250 to 400% of RP by tender. If the minimum price is not fetched, the plot is not being sold. Land sold for social, cultural, religious or educational purposes is subsidised.[10] As per February 1995, land for religious purposes is being sold at around Rs 1,100 per m². For commercial purposes (plots larger than 20m²), CIDCO's land sales are always by tender, with a minimum of 600% of RP. However, the system has not always been like this. Earlier, commercial land used to be sold at a fixed price. From the moment the market picked up, however, that system was replaced by a sales system by tender, and still later increasingly by auction. The same procedure (by tender) has increasingly been followed in case of residential land sales. The tables and graphs hereunder shed some light on the sudden sharp increase since 1993 in prices received for land sold by tender.

Table 6.9 : Average rates received by CIDCO from land sales by tender in two nodes of New Bombay (CBD and Airoli), 1991-1995 (in Rs./m²)

	CBD			Airoli		
	Resid.	Shop + res.	Comm.	Resid.	Shop + res.	Comm.
90-91	7,606	7,896	NA	644	NA	NA
91-92	NA	6,186	NA	1,852	1,925	NA
92-93	6,126	8,623	9,394	2,232	NA	NA
93-94	7,676	2,000	NA	4,123	3,600	6,158
94-95	22,459	18,400	44,358	10,515	19,435	NA

Source: internal working document CIDCO (1995). Note: NA=not applicable (no plots sold by tender in given year)

Figure 6.8: Average rates received for residential land sales by tender, 1991-1995

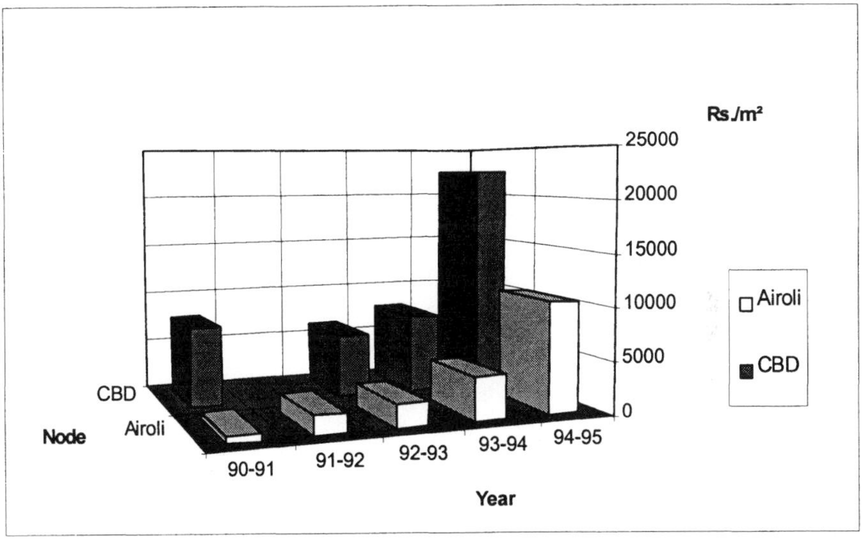

Based on figures table 6.9 (internal working document CIDCO, 1995)

Some other figures, not included here, suggest an even steeper increase in land prices for commercial use in CBD Belapur and Airoli nodes. Whichever figures are being considered, it is clear that in a short time span of some five years, commercial and residential land values in more developed nodes (like CBD) as well as in relatively less developed nodes (like Airoli) have increased manyfold. Moreover, CBD and Airoli are not exceptional nor extreme cases. A similar process can be observed in other nodes of New Bombay as well, with commercial land values per square metre currently (1995) at around Rs 60,000 in Vashi, Rs. 40,000 in Nerul, Rs. 35,000 in Koparkhairane, Rs. 20,000 in New Panvel, and Rs. 10,000 in Kalamboli (internal document CIDCO, 1995).

Looking at land sales patterns for residential, social and commercial purposes in just one node over a longer period, an even sharper increase in incomes from land sales can be observed. Whereas before the 1980s land in Vashi was being sold at Rs. 854,000 per hectare, in recent times the sales rate has tremendously increased to over Rs. 26 million per hectare!

Table 6.10: Land sales pattern in Vashi node (different sale systems), 1980-1995

	Area (in ha.)	Amount (in Rs.)	Rs. received / ha.
up to 1980	79.78	68,202,000	854,876
1980-85	108.48	294,042,000	2,710,564
1985-90	90.11	803,339,000	8,915,093
1990-95	46.46	1,227,601,000	26,422,740

Source: internal CIDCO document (20 Dec 1995). Note: includes sales of land for residential and commercial purposes, warehousing, schools, and other social facilities

Figure 6.9: Receipts from land sales in Vashi per hectare, 1980-1995

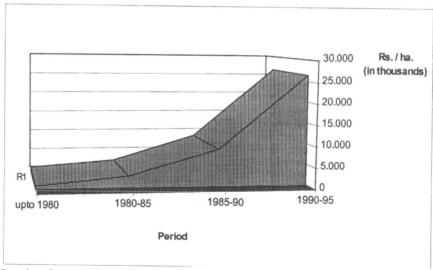

Based on figures table 6.10

Comparing the evolution in residential and commercial property rates in the private market in and outside of New Bombay, should expose possible geographical differences in land market activities. The figures in table 6.11 give the residential and commercial property rates in the private market in the three first developed nodes of New Bombay between July 1986 and April 1997.

Table 6.11: Property rates (built up) in three New Bombay nodes, 1986-1997 (in Rs. per ft²)

	Vashi		CBD		Nerul	
	Resid.	Comm.	Resid.	Comm.	Resid.	Comm.
07/86	300-600	600-800	250-300	450-800	275-300	450-700
10/90	500-700	1000-1500	500-600	650-900	400-500	600-800
06/91	600-900	900-1700	500-600	700-900	450-500	700-900
08/93	900-1800	2000-4000	525-750	1000-2500	550-750	1000-2000
07/94	1300-3000	2000-5000	750-900	1100-3000	700-1100	1100-2500
07/95	1400-3900	4500-8000	1800-3200	3000-7500	900-1600	2000-4000
03/96	1800-2500	3500-8000	900-2000	3000-6000	900-1500	2000-4000
04/97	1800-2800	2000-6000	1500-2200	3200-6000	1400-1600	2000-5000

Source: *The Accommodation Times*, 1986-1997. Note: *commercial* includes shopping; rates may be more by 50 to 200% depending on location, amenities, etc.

Figure 6.10: Residential property rates Vashi, CBD and Nerul 1986-1997

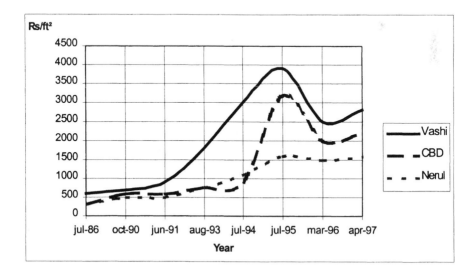

Figure 6.11: Commercial property rates Vashi, CBD and Nerul 1986-1997

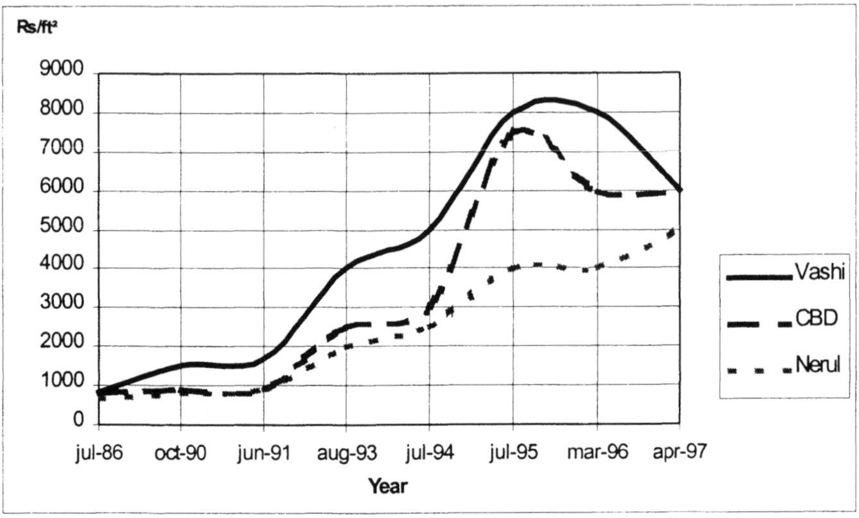

Based on figures table 6.11. Note: only the high rates have been selected

These figures and graphs clearly indicate the rapid increase in prices for property in New Bombay since the early 1990s, with a significant downturn (especially in the case of Vashi and CBD) during the last two years. Residential property rates in New Bombay have become almost as high as in Pune, another major city in the state of Maharashtra. Rates in Pune have also increased, but to a much lesser extent than they have in New Bombay. Commercial property rates have in New Bombay become even higher than in Pune.[11]

Two control groups outside of New Bombay were selected to test whether this steep increase is a specific characteristic of development in New Bombay alone. The first control group consists of three functionally distinct areas: Nariman Point, Pedder Road and Chembur in Greater Bombay;[12] the other control group consists of three administratively distinct areas: Thana, Kalyan and Virar outside of Greater Bombay in the adjacent northern districts.[13] The pattern in property rates in these areas can be observed from figures 6.12 to 6.15.

**Figure 6.12: Residential property rates in three selected areas of
Greater Bombay, 1986-1997**

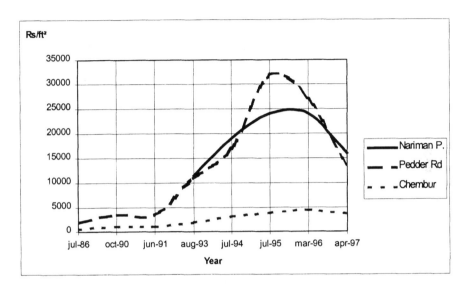

**Figure 6.13: Commercial property rates in three selected areas of
Greater Bombay, 1986-1997**

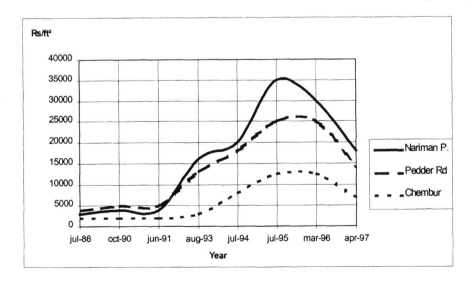

Figure 6.14: Residential property rates Thana, Kalyan and Virar 1986-1997

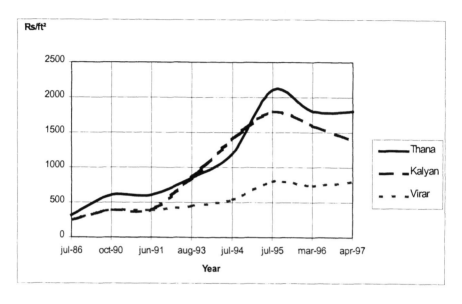

Figure 6.15: Commercial property rates Thana, Kalyan and Virar 1986-1997

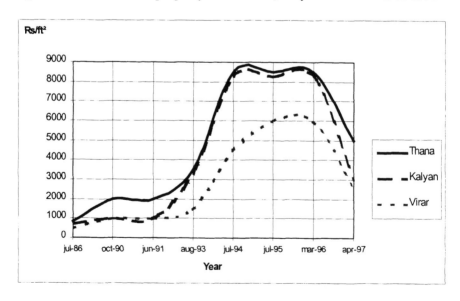

The graphs for Greater Bombay (figures 6.12 and 6.13) clearly show that both residential and commercial property rates have also tremendously increased in Greater Bombay between 1991 and 1995, after which a strong correction (an almost collapse) followed, especially in the two locations in South Bombay. In general, however, the overall pattern is fairly similar to the pattern in New Bombay, with that difference that the correction in New Bombay has been less dramatic than in Greater Bombay, and that the overall price levels obviously differ significantly.

The graphs for the three selected municipalities outside Bombay and New Bombay (figures 6.14 and 6.15) show a fairly similar general pattern of sharply increasing property rates in Thana, Kalyan and Virar between 1991 and 1995. The overall pattern in the three municipalities only differs in so far as it shows a fairly moderate decrease in residential property rates (and a status quo in the case of Virar).

Overall, it can be concluded that the sharp increase in property rates has not been a distinct characteristic of New Bombay alone. This pattern can indeed be observed in the entire Bombay Metropolitan Region. The major factor behind the dramatic increase since the early 1990s is, most probably, the liberalisation of the national economy with more flexible investment regulations since 1991, with Bombay increasingly becoming the focus of business and private capital, and the city's land and housing markets increasingly becoming the focus of a speculative craze. The fall in price levels since 1995 is a clear correction to the artificially high (speculation-induced) rates of the previous period.[14]

Impact of increasing land prices on further development   Because throughout the 1970s the value of the land was low, the financial returns from the sales of land were rather insignificant. Consequently, development activity in general was sluggish and not very diversified, not in geographical terms nor in terms of infrastructure and services. Public housing construction was to a large extent financed by loans from the Housing and Urban Development Corporation (HUDCO), and for a number of specific projects from the World Bank. Hence CIDCO's capital expenditure before 1980 was very much restricted, and cumulative expenditure up and till then did not exceed the Rs 400 million mark (Gill *et al.* 1995).

All this was obviously to change dramatically from the moment the land market picked-up (figure 6.16). Following certain positive developments (primarily the shift of the first wholesale markets into New

Bombay and the construction of a railway network) the demand for land and housing suddenly increased by the late 1980s, the markets reacted immediately to the sudden spurt, and incomes from land sales and from housing sharply increased (as is indicated by table 6.10 and figures 6.10 and 6.11). Moreover, with the increase in population from the 1990s, the receipts from individual 'service charges' and from general taxes (which until recently were exclusively levied by the NTDA) added to the increased budget. Consequently, this increased budgetary strength made it possible, from the late 1980s, to commence a more diversified development in several nodes at the same time, and to further invest in the road and railway network.

Figure 6.16: Expenditure and receipts of CIDCO 1976-1996 (in million Rs.)

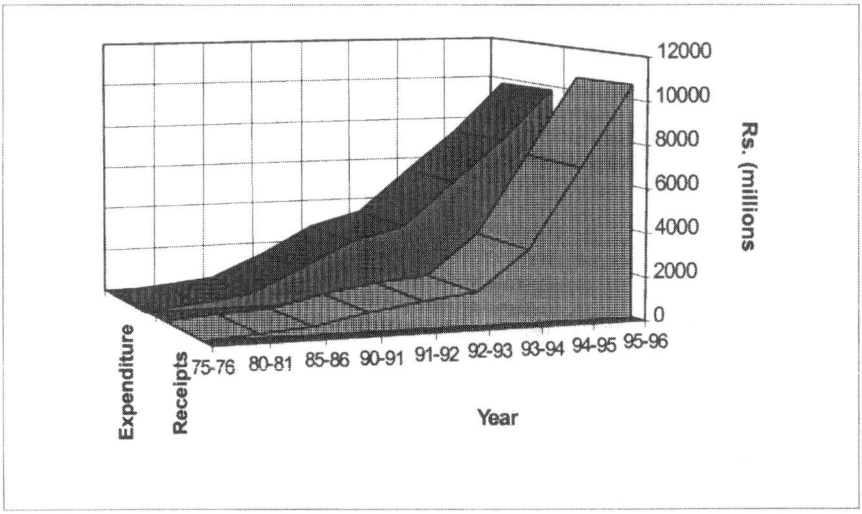

Source: figures from CIDCO departments and Gill *et al.* (1995). Note: 1976-1991 five-year periods, since 1991 annually

Government activity

An important aspect of the NBDP and one of the proposed major catalysts for initial growth in New Bombay, was the proposal to shift a major part of the state government's political activities out of South Bombay and into New Bombay's CBD. At present, only three sectors (sectors 10, 11 and 15) in New Bombay's CBD in and around Belapur node have been developed for the purpose of commercial and administrative functions, whereby the major part of government activity is to be found in sector 10 and whereby sector 15 remains largely untouched.[15] The total area covering these three sectors is 61 ha of land, at FSI of 1.5. This means that there is 91.61 ha of office space available, which is sufficient to accommodate 100,000 commercial and administrative jobs. Up to January 1995, 36.83 ha of land (or 55.24 ha of office space) had been sold, of which around 27 ha has been built up.

Table 6.12: Land sales in the CBD of New Bombay, 1970-1995

	Land sale (in ha)	Office space (in ha)
Up to 1980	2.90	4.35
1981-1990	28.66	42.99
1991-1995 (Jan)	5.27	7.90
Total	36.83	55.24

Source: Project Report CBD (1995). Note: 1991-95 figure is 5-year period

However, whereas well over half of the land of the entire CBD at Belapur has been sold, and 75% of that land has been built-up, the figures might give a wrong impression of activity in the new CBD. First, the major part of the plots have been sold to large building corporations and other private companies, but although two of the three sectors (10 and 11) are almost completely covered with mid-size office buildings, very few of them (especially in sector 11) are actually occupied (photograph 6.1).

Secondly, the number of government offices in the new CBD is also relatively small, and consists almost exclusively of newly created government offices (NMMC, CIDCO Bhavan, Konkan Bhavan and Raigad Bhavan) which have for obvious reasons been located in New Bombay.[16] Figures of government offices that have been shifted from Bombay's CBD at Nariman Point are not available, but it is quite clear that their number is insignificant. The currently 8,000-odd office jobs in New Bombay's CBD are therefore predominantly newly created government jobs, with the percentage of jobs related to private business restricted to 10% at most. The main cause for both phenomena, the absence of any major shift of government offices from Bombay to New Bombay, and the lack of any significant private business activity in the new CBD, has to do with state government neglect on the one hand, and with speculation on the other. We shall come back to this at a later stage.

Photograph 6.1: Largely completed but unoccupied commercial space in New Bombay's CBD at Belapur

Transportation

When the New Bombay project took off in the 1970s, several major roads such as the national highway (NH) No. 4, linking Bombay to Pune, were already crossing the area, and the only connection between Bombay and New Bombay was the old bridge in the northern town of Thana. At present, New Bombay is served by this older highway in the eastern section of the area, by the Sion-Panvel Expressway in east-west direction, and by the Thana-Belapur road which runs north to south over a distance of about 25 kilometres, and which divides CIDCO's development area from the MIDC industrial strip. A fourth major road (a new NH) is to be constructed between Kalamboli in the east and the JNPT port near Uran in the south-west over a distance of 27 kilometres.

In those days, bus transportation was the only public transport facility available. Public bus services were exclusively delivered by the Bombay Metropolitan Transport Corporation (BMTC), CIDCO's own bus company. In 1983-84 this company stopped functioning after a major conflict with the unions (the kink in Figure 6.17), and services were being taken over by both the BEST (Greater Bombay's municipal bus company) and the ST (State Transportation). From January 1996, the newly established New Bombay municipal authorities started their own bus service, the NMMT.[17] The figure shows the increase in the number of buses that operate in the new town since 1973.[18]

Figure 6.17: Total number of buses operating in New Bombay 1973-1996

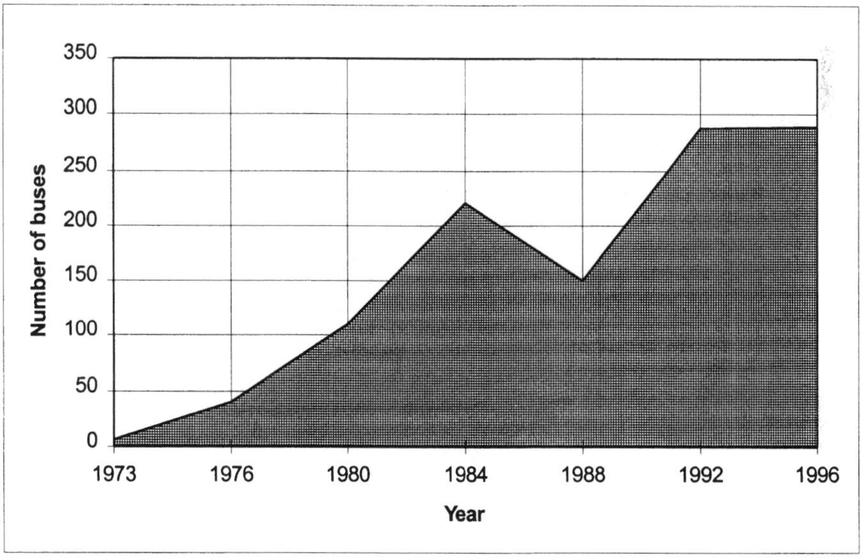

Figures collected from transportation departments CIDCO, BEST and NMMC. Notes: 1973, 1976 and 1980 only BMTC; BMTC stops functioning in 1984 and from now BEST and ST servicing the new town; 1988 and 1992 figures are estimates; first operation NMMT in 1996.

The idea to construct a bridge connecting New Bombay in Vashi with the eastern suburbs of Greater Bombay at Mankhurd had already been commissioned at the time of conception of New Bombay. This first Thana Creek bridge started operating soon afterwards (1972), servicing both public (buses) and private traffic (cars, trucks, motorcycles). Since November 1994, a new bridge at the same location has partly been constructed and is in operation. When completed, the Thana Creek bridge will have eight lanes for motorised traffic. The total cost of construction of the bridge has been estimated at nearly Rs. 100 crores (1 billion); funds which were to be recovered from toll charges on every vehicle (private *and* public) using the bridge. Figure 6.18 gives an indication of the increase in traffic over the bridge over time.

Table 6.13 : Number of motorised vehicles crossing Thana Creek Bridge, 1976-1996 (average per day)

Month/Year	Cars	Buses	Trucks	Total
12/76	2,340	1,414	2,365	6,119
12/81	6,072	2,862	5,118	14,052
05/85	7,196	3,110	5,769	16,075
05/87	10,504	3,413	8,296	22,213
12/92	12,621	5,120	11,631	29,372
05/95	19,721	6,052	16,371	42,144

Sources: 1976 to 1987 figures from Transport and Communication Division BMRDA, 'Basic Transport and Communications Statistics for BMR' (March 1991); 1992 and 1995 figures from Public Works Department, Government of Maharashtra. Note: the number of motorised traffic is counted for 7 days in a row, on the basis of which an average per day is calculated. This survey has been done twice a year, but not on a regular basis, since 1976.

Figure 6.18: Number of motor vehicles crossing Thana Creek bridge, 1976-1996

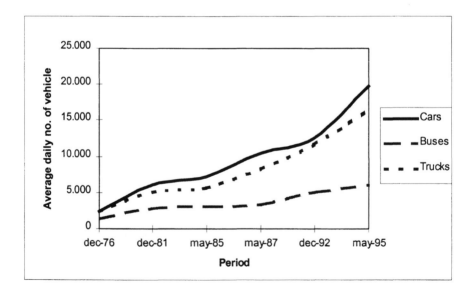

Recently, work has started and is in progress at Airoli to construct a second bridge connecting New Bombay with the Bombay Suburbs at Mulund. Plans for a third bridge (or tunnel-cum-road bridge), directly connecting the southern Uran area with the major job centres of South Bombay are sometimes being discussed, but so far the cost of around Rs. 1,400 crores (14 billion) is a major obstacle. It is believed by insiders that this trans-harbour link is not going to be sanctioned within the first six years.[19]

In the early 1970s, the NBDP proposed that when a sufficient number of people would be living in the area and bus transportation would become too expensive, a railway network would be constructed which would then become the major mode of mass transport in the area. Construction of such a railway network was started in the late 1980s, in a first phase connecting Vashi by bridge with the Greater Bombay eastern suburbs, and from there via Kurla into the CBD of South Bombay. Since then, the network was gradually extended and is planned to become a network of six rail corridors running north to south and east to west, totalling 157 kilometres in length. In May 1992, the first part of the major link between Mankhurd in Greater Bombay and Panvel in New Bombay, via the new Thana Creek railway bridge, was officially opened. At present, the entire link up to Panvel (via Vashi, CBD Belapur, Kharghar and Khandeshwar) is completed and has been in operation for some time.[20] Construction of a goods railway link between Kalwa and Nerul has been completed but is not operating yet. Construction of the railway line connecting Belapur with Uran in the deep south of New Bombay is about to start.[21] All the other major links are yet to be built or are in the initial phase of construction. Overall, roughly 20 to 25% of the network has been constructed.

Financing of the New Bombay railway network is a joint effort of CIDCO and the Indian Railways (central government).[22] Part of the repayment of CIDCO's investments is to come from the exploitation of the commercial space that has been built above the railway stations (which are exceptionally modern constructions to Indian standards), and it is the first time in India that a commercial exploitation of this kind is being planned.[23] The remaining part of the repayment is to come from overall land development and appreciation of land prices, and from the collection of a surcharge on railway tickets. Figure 6.20 gives an indication of the overall success of the railway system. It shows the sharp increase of both the number of tickets sold at Vashi station (the first train station of New

Bombay that started operating in 1992) and the total amount of money received from sales in that station.

Figure 6.19: Number of card tickets sold and receipts from sales at Vashi station, May 1992-December 1995

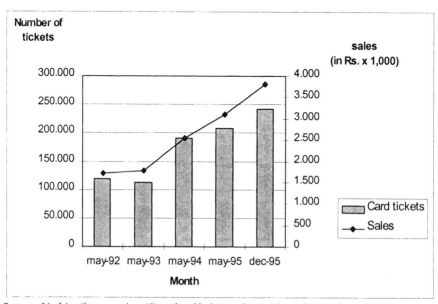

Source: Vashi railway station (*Details of balance sheet*). Note: the number of tickets and of sales is for one month only. The total amount received from May 1992 until January 1996, from both card tickets and season tickets, was Rs. 115.17 million. Notice the 5-month period between May and December 1995, as compared to the 12-month periods earlier.

Apart from mass public transport (buses and trains), a number of privatised modes of transportation have also developed. Within every node, rickshaws are the major means of service transportation, and every node is having several taxi stands. The NTDA also developed the infrastructure for India's first hovercraft service, which started operating between Vashi and 'The Gateway of India' in South Bombay in May 1994, and a second one between CBD Belapur and the Gateway in January 1995. A number of surveys conducted by the NTDA had indicated that 15,000 to 20,000 people would be using water transport by the turn of the century, which according to the NTDA would mean a substraction of 7,000 to 8,000 vehicles on the road. These hovercraft services are privately operated and have been developed mainly for business purposes.[24] Also planned for business purposes is the New Bombay 'City Aerodrome', a new local airport that is planned at Panvel to meet short haul air travel demand. So far, however, construction has not yet started. Finally, discussions have been going on for some time on a new major international airport just south of New Bombay at Alibag.

Social provisions

A final set of data that have been collected on growth and development in New Bombay are those in relation to education, health and community facilities. The significant increase in the total number of primary and secondary schools and colleges in New Bombay nodes (provided by the NTDA), and in the total number of students over the years can be observed from table 6.14.

Table 6.14: Number of schools provided by CIDCO in planned nodes and number of students

	Primary schools	Composite schools	Degree colleges	Professional colleges	Total no. of students
1975-80	NA	4	1	NA	21,518
1980-85	11	9	2	1	43,036
1985-90	30	24	13	5	64,554
1990-95	51	42	18	15	86,072

Source: social services department CIDCO. Notes: several sets of data were provided, every single one of them different from the other, which means that the validity of some figures may be questioned; education in the villages and in the municipalities of Panvel and Uran are not included in this table.

Health facilities in New Bombay nodes were to be provided according to the rules set by the state government, viz. two hospital beds per 1,000 population for a general hospital, and one bed per 1,000 for maternity. Therefore, given the current population figure of 700,000 the total number of beds in New Bombay at present is to be 2,100 (1,400 + 700). According to official NTDA figures, that number is with 1,952 slightly below the target. Yet, only a limited number of hospitals and dispensaries are fully operational, and the number of beds actually available is far below the official figures given. None of these medical facilities are managed by CIDCO. The NTDA's only involvement is in providing either the building, or the plot of land. Apart from these hospital beds, the NMMC is currently developing 11 Urban Health Posts for preventive and curative health services at primary level.

Table 6.15: Health facilities (number of hospital beds) in New Bombay nodes, 1975-1995

	CIDCO	NMMC	By trust	Private (a)	Total beds
- 1985	80	-----	50	NA	130
1990	355	-----	-----	NA	
1995	530 (b)	200 (c)	600 (d)	622	1,952

Source: public health dept. CIDCO. Notes: (a) consists of approximately 50 private nursing homes, but no breakdown per period available (NA=not available); (b) one dispensary (25 beds) not yet operating; (c) two dispensaries (50 beds) not yet operating; (d) only some 150 beds operational.

The increase in the number of social, cultural and religious institutions in New Bombay has also been fairly rapid. The NTDA has itself provided some social facilities (such as community centres), whereas other institutions were given land at concessional rates. Currently, a plethora of socio-cultural and religious institutions can be found in the nodes, ranging from sports complexes to homes for the aged, temples, churches, mosques, *gurudwaras*, recreational centres, gymnasia, etc... In February 1996, only the number of social welfare institutions in New Bombay was no less than 74.

Conclusion

Although real development is only a recent phenomenon (dating from the late 1980s and early 1990s), and although the success has partly been the result of external factors, in terms of growth and development in general the New Bombay project must be considered as very successful. Between 1981 and 1991 the population of the twin city more than doubled to 535,000, and at present has already crossed the 700,000 limit. Further, industrial and commercial activity also rapidly increased since the late 1980s, an extensive transportation network was developed with traffic movement constantly increasing, and the number of social, cultural and religious organisations are all indications of a growing urban dynamic.

The rapid growth and development of New Bombay since the late 1980s is reflected by the increase in public and private construction activity, as well as in the very sharp increase in residential and commercial land prices and property rates since 1991. This pattern, however, is not a distinct feature of growth and development in New Bombay alone, as property rates in other areas of the BMR have followed roughly the same path. What is important is that the sharp increase in land prices in New Bombay had a tremendous impact on the overall development pattern in the region, as the NTDA's working budget was augmented manifold.

To the question why growth and development in New Bombay took such a sudden fast pace from the late 1980s and early 1990s, no single answer can be put forward. In the early 1980s several minor factors have stimulated initial growth: the level of compensation money for the acquisition of land was augmented in 1983-84, which gradually activated the demand for housing with public financing agencies coming forward to meet that demand. Later, by the early 1990s, the construction of a first major railway line and bridge, linking New Bombay with the job centres of Bombay, on the one hand, and the liberalisation of India's economy since 1991 on the other, were two factors that had a tremendous impact on developments in New Bombay. Furthermore, the commissioning of a second Thana Creek bridge at Airoli, the streamlining of New Bombay's telecommunication system with that of Bombay's, implementation of the plan to shift part of the wholesale APM to New Bombay, and the commissioning of a high-tech port near Uran, significantly added to the sudden growth in New Bombay since the late 1980s.

If one is to speak of a healthy, balanced and sustainable urban development, it is however not sufficient to turn our focus exclusively on aspects of physical and economic growth, and on such things as financial mechanisms and budgetary viability. If a large rural area of the size of New Bombay is to be developed as an important new urban area, one of the most obvious and indeed most important questions that has to be addressed is to what extent the development of the area has had an impact, be it positive or negative, on the lives of the people that have traditionally inhabited the area. This is the question that will be put forward in the next chapter.

Notes

[1] The Directorate of Industries has, in the light of the state's industrial location policies, to approve all investment proposals in the state, which is why these data are available per District; the number of State Districts has been augmented from 26 to 30 in 1991, and the number of Divisions (each grouping a number of Districts) from 4 to 6 in the late 1980s.

[2] Investment in the Ratnagiri District is slightly higher than in the Thana District, but is almost exclusively the result of investment by only four large companies (of which the Dabhol Power Company, i.e. Enron, and the Hindustan Oman Petroleum Ratnagiri Corporation take the largest share with Rs. 9053 crores and Rs. 4,500 crores respectively).

[3] The only exception is the Nagpur Division, where investment is slightly higher than in the Thana District (Rs. 18,532 crores as compared to Rs. 17,210 crores), but given the high number of undecided investments in Thana and Raigad, the total amount of both will be far higher.

[4] No comparable figures are available for industry and commercial activities in the 1980s separately. A breakdown of current commercial activities in New Bombay would show that retail trade, the wholesale markets and the public sector offices employ over half of all workers in the commercial sector.

[5] Some of these data slightly differ from other sets of data.

[6] This initiative was made possible under the Maharashtra Agriculture Produce Marketing (Regulation) Act 1963, which enables the government to denotify existing markets and shift them to other locations.

[7] Some of the developments in progress (or in the pipeline) are the expansion of the terminal infrastructure, the development of a marine chemical terminal, the expansion of infrastructure facilities, the development of recreational facilities, the development of an export processing zone, construction of a ship repair unit, construction of a private container freight station, and a plan for development of ro-ro facilities.

[8] As ports are an integral part of the central government's responsibility, the state government nor the NTDA have anything to do with developments in the JNPT. Both have only been involved in the acquisition of land and in the provision of basic infrastructure in the area.

[9] As has been explained earlier, the RP is a fixed price which can be considered as the actual cost price, i.e. all the development costs of a node divided by the available saleable area in that node.

[10] Primary schools pay nothing; secondary schools pay 10% of RP; higher education 50% of RP; social, religious, etc... pay also 50% of RP; and land for public utilities (telephone, mail, electricity company,...) is being sold at 75% of RP.

[11] In March 1996, top residential property rates in Pune were at prime locations between Rs. 1500/ft² (Laxmi Rd.) and Rs. 3500/ft² (Bundh Garden, Koregaon Park, Pune Camp); top commercial rates were between Rs. 3000/ft² (Laxmi Rd.) and Rs. 4500/ft² (Bundh Garden, Koregaon Park, F.C. Road), with an exceptional Rs. 12,500/ft² in Pune Camp.

[12] Nariman Point is the old CBD of South Bombay in which there is only recently some residential activity, Pedder Road is a posh location in the northern part of South Bombay with a mixture of residential and commercial functions, and Chembur is a lower-middle class and industrial location in the eastern suburbs of Greater Bombay.

[13] Thana is an old and large municipality directly north of Greater Bombay, Kalyan is a fairly new satellite town north-east of Greater Bombay and New Bombay, and Virar is a fairly small and semi-rural township further north of Bombay and Thana (and since recently also under the development authority of CIDCO).

[14] At present, several business houses are actually moving out of Bombay (for instance to Bangalore) to escape the sky-rocketing property rates. This move may have an impact on the fall in price levels since 1995.

[15] In March 1996, the number of government related offices in sector 10 was 60, employing 4,622 persons.

[16] At present, about 2000 and 3000 people are employed in CIDCO Bhavan and Konkan Bhavan respectively (the buildings of the NTDA and the state government), with an additional number in the Belapur Bhavan (NMMC offices) and the Raigad Bhavan (District authorities).

[17] The ST buses operate mainly on the major roads running through New Bombay, excluding any connections with Greater Bombay, the NMMT and BEST buses operate within and between the nodes, with the latter connecting New Bombay with locations in Greater Bombay.

[18] As no exact figures on bus transportation in New Bombay are available for the entire period, the data have been brought together from different sources and are sometimes 'official' estimates.

[19] From a talk with K.T. Lele, Additional Chief Engineer Transportation and Communication Department CIDCO.

[20] CBD Belapur station has been functioning since '94, the extension to Kharghar and Kandeshwar was officially inaugurated in Jan '95. Total cost of the link between Mankhurd and Belapur was estimated at Rs. 287 crores.

[21] Early 1996 it was believed that financing of this link was going to be included in the new budget of the central government of March 1996, after which construction could start.

[22] Two-third of the Mankhurd-Belapur line has been financed by CIDCO and one-third by the Railways; the same financial construction is to be adopted for the Belapur-Uran link. The goods railway line between Kalwa and Turbhe has been financed by CIDCO and the BMRDA in a 50-50% partnership.

[23] Vashi and CBD Belapur, for example, are together providing 130,000 m² of office space above their railway stations. Construction of these offices is, however, still in progress.

[24] In the meantime, the link CBD Belapur-Gateway is no longer operating due to a lack of public interest.

7 Impact of Growth of New Bombay on Original Population

'The Corporation as its prime duty looks after the rehabilitation of project affected persons...'

(CIDCO, *Two Decades of Planning and Development*)

Introduction

When the development plan was published for public comments, one of the five broad objectives in compliance with the main objectives of the project by the state government, was 'to provide training and all possible facilities to the existing local population in the project area, in order to enable them to adapt to the new urban setting and to participate fully and actively in the economic and social life of the new city' (NBDP 1973:10). With this objective, the NTDA was thus also given the social task of rehabilitation and integration of the local population and the villages into the new urban fabric; a task of utmost importance in terms of social justice and welfare as well as of geographical physical and social balance.

As we have seen earlier, about 56% (or 166 km²) of the total net New Bombay area of 294 km² was private land, distributed over a large number of people and used mainly for the cultivation of rice and other crops by the majority of the 120,000-odd local population spread over the 95 villages located in the project area (figure 7.1). Apart from the 34% of land that was already owned by the government, all this private land was to be acquired by the state government under the Land Acquisition Act, and to be handed over to the NTDA, so that ultimately all the land in the area would be publicly owned. As this land was to function as the main resource for development of the area, this large-scale acquisition was obviously a necessity and the first major step to be taken. Consequently, the local landowners would lose all their land, and with that also their income and life sustenance. For that loss they were to be financially compensated.

171

Figure 7.1: The 95 original villages of New Bombay

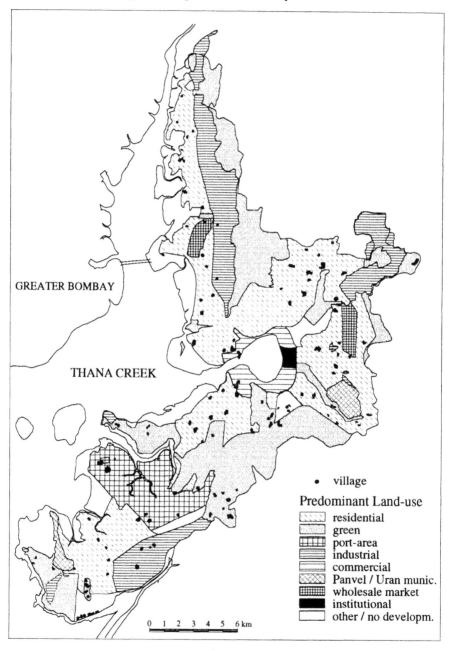

In this chapter we will discuss to what extent the NTDA has been successful in implementing this social task of rehabilitation and integration of the local population. How did the process of acquisition of land proceed over the years, and how was it compensated? Which programmes have been introduced by the development agency in terms of rehabilitation and integration? And, most importantly, how successful have these programmes been in practice and to what extent has the local population been affected by the implementation of the New Bombay project?

In relation to these questions, two different sets of data have been collected: a first set was based on secondary data, a second set on empirical research. First, information in relation to the acquisition of land, and to the planned rehabilitation programmes and integration procedures has been based on data which have either been published in reports, or have been collected through the NTDA's Rehabilitation and Social Services Department. In addition, some rough data have also been taken from a large socio-economic questionnaire survey in the villages, which at the time of my fieldwork was being conducted by the NTDA in 54 villages (25% sample, including nearly 49,000 people). Data taken from this survey will be indicated as *Village SES-96*. Finally, all data in relation to the implementation of these programmes in practice, and the concrete impact of the New Bombay project on the original population, has been based on a fairly extensive personal village survey by using Rapid Rural Appraisal (RRA), and to a minor extent Participatory Rural Appraisal (PRA).

RRA and PRA are survey techniques based on interviewing, which is only one of the possible research methods available in quantitative and qualitative social science research. Because RRA and PRA are fairly new techniques and are not widely known or used, we have, in order to clarify the place and functions of RRA within the social science methodological field, included a brief discussion on methodology, viz. on the ongoing debate between quantitative and qualitative research, on the research method of interviewing in general, and on the concept, the choice, the practical usefulness and the accuracy of RRA and PRA research. The discussion has been included as an appendix.

Land acquisition and rehabilitation measures

Since the initial years of development in the early 1970s, the process of acquisition of private land in New Bombay has been very slow and painful, and has been a major problem for everyone involved. Even today, after 25 years of project implementation, still over 4,000 ha (more than 20%) is to be acquired, and numerous law cases between landowners and the government are pending with the courts.[1] A direct consequence of the stiff opposition of the landholders was that in 1984 the Land Acquisition Act was amended. It stipulated that all the lands which were involved in on-going cases would have to be acquired within two years or that otherwise the notification for these lands would lapse. Consequently, within the next two years, the most urgently needed lands were acquired, but notification of more than 5,000 ha. of land lapsed. For these lands, the notification process had to start all over again, and more delays were lying ahead.

Apart from the financial compensation, the NTDA introduced several rehabilitation measures for the Project Affected People (PAPs):

(1) A first series of measures were focusing on the villagers individually, and foresaw in college stipends for the children, in technical training schemes, in alternative employment opportunities in public and private companies, in cheap loans to petty traders, in allotment of shops and stalls, and in small contract works.

(2) A second series of rehabilitation measures were focusing on the villages within the project area internally, which according to the Development Plan were physically not to be adversely affected by development of the nodes. In these villages, the NTDA was to provide or to finance a minimum of physical and social infrastructure such as water supply, better sanitary facilities, a minimum number of toilet blocks, a village tank, construction or improvement of approach roads, construction of a storm water drain where needed, a village 'Samaj Mandir' (temple), construction or upgradation of primary village schools, and the provision of a primary health centre.

(3) A third series of rehabilitation measures for the PAPs had to do with the ownership of land. In the first phase (in 1978) the NTDA introduced the so-called Gaothan Expansion Scheme (GES). Under this scheme, the erstwhile landholders whose lands were acquired, were provided with residential plots in a layout prepared on the periphery of the existing villages, mainly in order to accommodate the natural increase of the village population. Ten percent of the total acquired land of the village was to be reserved for this scheme, half of which was to be used for housing purposes for the PAPs, half for the provision of facilities. Every landholder that had lost at least 100 m² of land was entitled to a plot of 5% of his former plot of land, with a minimum of 100 m² and a

maximum of 500 m². The lease premium charged for such plots was double the average cost of acquisition plus a nominal charge of Rs. 5 per m² as land development cost. Those who had a plot of less than 100 m², or had no land at all, were to be given a plot of 40 m². Finally, a number of restrictions were introduced to avoid easy transfer of these lands.

For those landholders who had previously owned more than 1 ha. of land, this 500 m² ceiling was a major source of opposition to the GES. Moreover, all the former landholders, well aware of the large amounts of money the NTDA was fetching from the sale of their former lands since the mid-1980s, demanded some share in the enhanced value of the land. Opposition and resentment were stiff, and in 1986 the state government finally amended the entire GES scheme. From now, each landowner was to be given back 12.5% of his acquired land at FSI 0.75, without an upper limit, excluding such categories as absentee landlords, trust and company landholders, or proprietors of salt pan lands. Of the land so recovered, 15% of the FSI was allowed to be used for commercial purposes. The price to be paid to get this 12.5% of developed land back was not changed, and remained twice the acquisition cost plus Rs. 5 per m² as land development cost. Whereas initially this scheme was applicable only to those whose land had been acquired before February 1986, continuous pressure from the PAPs finally led to the extension of the scheme to all landholders. Nevertheless, whereas the FSI was later increased to 1.5, it would take until late 1994 or early 1995 before the then state government (led by Congress leader Sharad Pawar) gave its approval to start releasing lands under the 12.5% scheme. It was calculated that 10,300 families would be gaining from this scheme. From the official side it was, therefore, expected that this scheme would put an end to the 24 year old opposition of the villagers and to the rehabilitation problem of New Bombay's PAPs. By that time, the expenses on rehabilitation of the PAPs had increased substantially, from nearly Rs. 12 million in 1990-91 to an estimated Rs. 185 million in 1994-95. Tables 7.1 to 7.3 are based on the NTDA's official figures and give an indication of the number of individual households and villages that have benefited from each scheme. Figure 7.2 gives the amount of money that has been spent on rehabilitation.

Table 7.1 : Household rehabilitation programmes and number of PAP Beneficiaries, 1994

Programme	Beneficiaries	Programme	Beneficiaries
college stipends	4,661	boys placed (CIDCO + elsewhere)	4,469
technical training boys	3,445	Shops	202
mali training	171	stalls and otlas (a)	28
auto rickshaw drivers	259	loans to petty traders	85
electricians	90	quarry permissions	244
plumbers	30	contract works	1,079
carpenters	63		

Source: Rehabilitation and Social Services Department CIDCO. Note: several series of data have been collected which were slightly different from one another; (a) *otla*: small platform for commercial activity (*shop* is the largest, *stall* is smaller, *otla* is smallest)

Table 7.2: Village rehabilitation programmes and number of villages that benefited, 1994

Programme	Villages	Programme	Villages
grant for college building	1	library / books	1
grant for secondary school	8	gymnasium	1
grant for primary school	70	crematorium	19
community centre	10	repairs to bund	9
water supply	46	dhobi ghat (a)	5
toilet blocks	46	cattle pond	1
urinals	36	t.v. sets	47
village Panchayat office	3	adivasi houses (b)	1
S.T. bus stand	1	stage	1
approach roads	28	playground	2
gutters	10	primary health centre	1
lake	2		

Source: Rehabilitation and Social Services Department CIDCO. Notes: (a) *dhobi ghat* = place for washing clothes; (b) *adivasis* = tribal communities

Table 7.3 : Villages under GES or 12.5% land compensation scheme, 1994

Gaothan Expansion Scheme	9 villages
12.5% scheme	52 villages (proposed)

Source: Rehabilitation and Social Services Department CIDCO

Figure 7.2: Expenses on rehabilitation per year up to 1994-1995

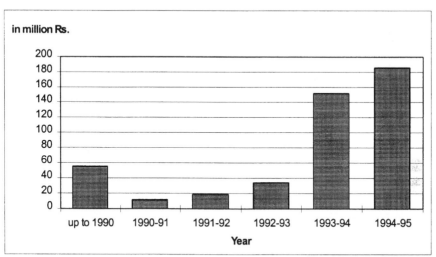

Based on figures from Rehabilitation and Social Services Department, CIDCO. Note: 1994-95 figure is provisional; Total expenses up to March 1994 was Rs. 27.22 crores (272.2 million)

The village survey

Sample construction

As has been said in the introduction, the village survey was conducted on the basis of RRA, and to a lesser extent PRA. Within a period of 22 days, between 18 April and 2 May 1995 and again between 30 November 1995 and 10 January 1996, 78 respondents in 18 different villages were interviewed (individually or in group), with a minimum of three in each village. The interviews were taken during day-time as well as in the evening, and during the week as well as in the weekend. The villages were selected in a way that would guarantee a geographical as well as a functional balance.[2] In the developed northern part of the new town, 12 villages were selected and 49 interviews were taken. In the less developed southern part another 6 villages were selected and 29 interviews were taken. In both areas, north and south, the selected villages were geographically spread. Some villages were selected for specific reasons: Vashi gaothan because it used to be predominantly a fishing village;

Turbhe gaothan because it is the nearest village to the highly developed Vashi node; New Sheva because it is a village that was completely shifted from its original location in the JNPT port area. The villages that were selected were Belapur gaothan, Karave, Diwala, Sanpada gaothan, Airoli gaothan, Rabale, Turbhe gaothan, Vashi gaothan, Khairane gaothan, Shirona, Nerul gaothan, and Kukshet in the northern part, and New Sheva, Bokad Vira, Panje, Pagote, Karalpada, and Sonari in the southern part of the area (figure 7.3).

The rough data of CIDCO's *Village SES-96* give an indication of the social profile of the population in the New Bombay villages. The male population is slightly larger (53.3%) than the female population (46.7%). Households of 4 to 5 members are to be found most frequently (47.4%), whereas 25.6% of the households have 6 to 7 members, and 14.3% have 2 to 3 members. Just over 94% of the population is Hindu, and the language spoken by the majority is the local Marathi language (89.9%).

The respondents in the villages (sometimes households, sometimes individuals, occasionally fairly large groups of villagers) were randomly selected. An effort was made to include the village 'Sarpanj' (village head) into the sample in every village that was visited. In the northern part of New Bombay, where over the years the NTDA has gradually taken over all the municipal functions from the former local councils, the Sarpanj has lost all his former responsibilities but is still a person very well placed to talk about the changes in the village over time. In the southern part of New Bombay, on the other hand, where development of the land was only recently started, the Sarpanj is still the official head of the village. In 8 of the 18 villages that were visited we have been able to interview the village Sarpanj. In several villages we have also been taken around on so-called 'transect walks', mostly by a local social worker, sometimes by some other person simply interested in what we were doing.[3] Finally, as most PAPs since the early days of the New Bombay project were organised in the so-called 'Khari-Kalwa Belapur Vibhag Shetkari Samaj Sanghatana', we also had an interview with P.C. Patil, a local advocate and leader of the PAP-organisation.

Each interview, depending on the quality of the respondent(s), has taken between 45 minutes minimum and three hours maximum. All the interviews were taken by myself with the help of an English-Hindi-Marathi translator. Two different persons were selected as translators, both of them males;[4] one was a postgraduate student in economics, the other a local

Figure 7.3: The 18 villages that were covered in the survey

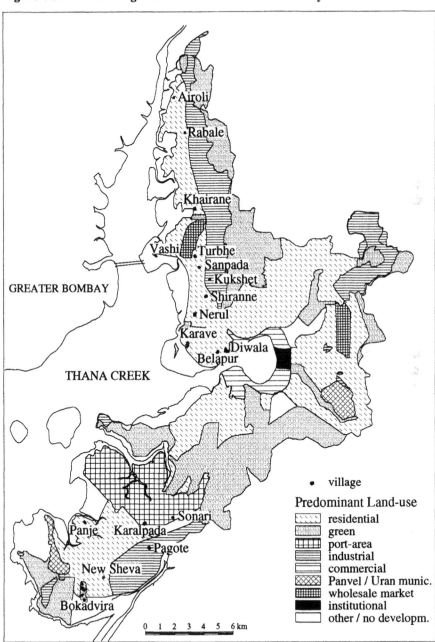

Baptist pastor. Both were provided with a basic knowledge of what RRA is about, had a very good understanding of my research intentions, and in the case of the pastor an excellent feeling for, and experience with approaching the local population. Given the group of respondents and possible feelings of hesitation and reluctance towards a 'white babu', the interviews were not recorded on tape (as this might have reinforced such feelings). Instead, detailed hand-written notes were kept.

At the end of each day, the fieldwork notes were typed into the computer and catalogued. For each interview, date, place and hour were registered; name(s) and age(s) of the respondent(s) was/were included, his/her/their profession, their possible family ties, the setting where the interview took place, and the general environment (the number of people actively or passively present at the setting, things that were happening around us and might have had an impact, etc.). For each interview, a number between 1 and 10 was given according to a respondent's overall motivation to answer to the questions properly and thoughtfully; a number between 1 and 4 was given according to our personal feeling of someone's overall capability to answer to the questions properly. Logically, in the final phase of processing and analysing the data, this kind of information was taken into account.

In terms of interviewing methods, a slightly provocative but gentle attitude was adopted. Wherever someone would be speaking very strongly in favour of the impact of New Bombay and of the NTDA's rehabilitation measures, we would mention some negative stories we were told earlier; wherever someone was speaking very strongly against New Bombay and the NTDA's activities, we would confront them for example with an official list provided by CIDCO of all the things the NTDA had already done to rehabilitate the local population and of all the facilities and infrastructure they had already provided in the villages. Such a slightly provocative attitude has helped in stimulating the motivation and in further deepening the respondent's arguments.

Impact of urban development

The quantity of data generated in RRA and PRA case studies can be extensive, and can pose certain problems when it comes to analysing the data and producing the major research results.[5] The main problems are those related to amount and focus: should everything documented be included, or should the production of data be limited and should it concentrate on the most important and most relevant generalisations? Stenhouse (1981) suggests that the answer to this question basically depends on whom the research results are meant for. He suggests that if institutions and individuals outside the immediate situation are the audience, generalisations might be more appropriate, with details serving to illustrate particular points. It is this approach that has been adopted in organising and analysing the fieldwork data.[6]

In the next sections, five topics will be discussed: (1) matters of land acquisition and financial compensation; (2) the extent of success of employment and other rehabilitation measures; (3) the general cost of living in the area; (4) the success of village rehabilitation and integration measures; and (5) the GES and the 12.5% scheme.

Matters of land acquisition and financial compensation   It has to be remembered that New Bombay consists of three distinct areas that fall under three different authorities: the industrial area between Belapur and Kalwa along the Thana-Belapur road is controlled by the MIDC, the area in and around the new port by the JNPT, and the rest of the area, the bulk of the land, is controlled by the NTDA, i.e. CIDCO. The areas under the MIDC and CIDCO thus fall under the Maharashtra state government, and the area under the JNPT under the central government. As a consequence, the financial compensation and rehabilitation packages for the affected villages and their population have been different according to the area in which the village is located.

From the late 1970s, the compensation that was paid by the NTDA for the agricultural land in most areas of New Bombay was put at Rs. 15,000 per acre (1,000 ft²), or about Rs. 37,000 per ha. (Rs. 3.7/m²).[7] Earlier, from the early 1960s, and long before CIDCO had moved into the area, the MIDC had already acquired the land on which the new TBIA (industrial belt) had to be established, for which it was to pay Rs. 5,000 per acre (Rs. 1.24/m²) for agricultural land and Rs. 2,000 per acre (Rs. 0.49/m²) for grass land. Finally, compensation that was to be paid by the JNPT for the acquisition of lands in the area which was to become the new port was put at Rs. 15,000 per acre (Rs. 3.7/m²) before 1984, and between Rs 22,000 (Rs 5.4/m²) and Rs. 27,000 per acre (Rs. 6.67/m²) afterwards. Land which was needed by the railways or by the electricity board has been acquired at different compensation rates. Finally, beside agricultural land, which was acquired in bulk, some people also lost their residential land.[8] According to the *Village SES-96*, of all the families that have lost their land, each household has on average been paid Rs. 4.3/m², and has received Rs. 37,342 in total.

As we explained above, the limited amount of compensation provoked a lot of opposition from the many small landholders. So far as acquisition of the MIDC lands in the 1960s was concerned, opposition was fairly limited, as every farming household was promised at least one alternative job in the new industries that were to come up in the MIDC area. Moreover, they were promised that no more lands would be required afterwards. However, some time later, when the state government came up with the New Bombay plan and much more acquisition of land was proposed, farmer's opposition obviously increased, the more since almost all the land was to be taken and the amount of compensation money that was proposed was considered as far too low. This is where most PAPs started to organise themselves. Those who could afford it took their case to court. Some of them have gained from it, but others, whose cases are still pending, have not received any money yet. Most former landowners however reluctantly accepted the money as they could not afford the judicial expenses, and thus did not have much choice.[9] Some of them received the compensation money on the spot, others have been waiting for years.

Employment and other rehabilitation measures   As mentioned above, in terms of rehabilitation there has been an important distinction according to the location of the village in CIDCO area, MIDC area or JNPT area. Another important distinction was made between the officially recognised PAPs and other affected village people. Indeed, not everyone affected by the New Bombay plan has been considered as a 'Project Affected Person', who was entitled to at least one alternative job and to a number of other rehabilitation measures. All those who were not possessing any land (those engaged in fishing activities, village artisans, tribals, and landless agricultural labourers or salt pan workers) have officially never been considered as PAPs. Thus, only the former landholders and their families were entitled to compensation and rehabilitation. Moreover, second generation PAPs (sons and daughters of a former landholder) who moved out of the parental house were no longer entitled to any benefits either, even if they were still living in the same village.[10]

The major factor of disappointment, opposition and anger of the local population towards the New Bombay project, is that of all the promises of alternative employment and other rehabilitation schemes very little has actually been implemented. At the time of acquisition, the MIDC, CIDCO and the JNPT all made bold promises regarding alternative employment. Official certificates were delivered to former landholders and were to assure them of a new job in industry or as contract workers. In the initial years, there seem to have been a number of genuine efforts to help the PAPs to survive within the uncertain and drastically changing environment. A significant number of villagers who lost their land (and job) were recruited in the new industries, or were given small jobs with the NTDA. Officials of the NTDA would visit the villages to find out the number of potential employees and would intervene with employers on their behalf.[11] As has been mentioned previously, CIDCO also ran its own bus company, in which some 1,800 people were employed, about half of which were PAPs. Moreover, the development agency also established a significant number of rehabilitation schemes that focused on education and technical training: children of PAPs could get grants from 10th standard onwards and could follow classes in one of the junior or senior colleges established by the NTDA, boys could get training in specially established technical institutes, and former landholders could also get technical training (as TV repairer, electrician, plumber, rickshaw driver or truck driver).

Unfortunately, with the years, the PAPs and their families seem to have been increasingly neglected, by the NTDA as well as by the private industries in the area. First, for village children it has become increasingly difficult to get proper training or education. A child that wants to get a college degree, will increasingly be confronted with financial and practical difficulties. Whereas primary education in the village schools is usually very cheap, the quality is mostly of a low standard. Therefore, he or she will be badly prepared to get a place at one of the high schools in the nodes. If a place at a high school is nevertheless secured, there will be no financial help from anyone until 10th standard. That most local households have increasingly been confronted with financial problems to meet the education bills is partly because of the NTDA's policies. Over the years, all the colleges and technical training institutes, which earlier had been built and controlled by the NTDA, have been privatised. As a result, 'donations' and tuition fees have been introduced and have risen constantly. At present, the donations (a kind of admission fees) may be anything between Rs. 2,000 and 10,000, according to the college (old or modern) and the medium (Marathi, Hindi, English). Such school donations have been formally banned by the state government, but in practice the system is still in full operation (only now slightly more hidden than before). The tuition fees, on the other hand, may be anything between Rs. 750 and Rs. 1800 per year.

Those who manage to finance such an adventure and pass 10th standard, are entitled to a CIDCO stipend of Rs. 1,500 to 2,000 per year, on the condition that the father is a genuine PAP and that the household's income does not exceed Rs. 10,000 per year. In practice, however, that stipend money may not even be sufficient to cover just the transportation expenses.[12]

A similar evolution and neglect could be observed concerning the NTDAs technical training institutes, which were established in order to teach the PAPs and their children some technical skills and give them a chance on the employment market. After a number of years these training institutes were also privatised, and a similar donation system has been introduced resulting in much higher expenses. Furthermore, simply reaching the NTDA training institutes, of which there were only a few, was far too expensive for most potential students. It is not surprising then that at both levels, technical training and college education, the number of drop-outs from the villages is very significant, and that the number of village students graduating from senior colleges is rather small.

Beside the increasing difficulties for the local youth in getting a proper education or training, an even far greater problem, for them and for all other affected persons for that matter, is employment. Even for those who manage to graduate from 10th or 12th standard or from a technical training institute, job prospects are far from good. The limited number of white collar jobs in the area is one problem, but in addition there is also a wide-spread reluctance of potential industrial employers, who consider the NTDA training programmes as qualitatively insufficient.[13]

The youth is only one group that suffers from insufficient opportunities on the job market. For the former farmers, fishermen and salt pan workers, things have also gone from bad to worse. The number of job rehabilitation schemes has been drastically reduced over the years, and overall interest in employment generation and economic integration of the village people seems to have become of minor importance. The NTDA's own bus company, which used to be a major employment base for the PAPs, stopped functioning in 1984, and all employees became unemployed. The number of households where one person (minimum) got an alternative job (as was promised again and again by NTDA, MIDC and JNPT) is relatively small, the more since the local population has also increasingly been bypassed for employment by the private sector, and more and more outsiders have been recruited. The official reason for the latter phenomenon given by the employers is that the villagers are very traditional and religious people, who celebrate many religious festivals, due to which a lot of absenteeism is prevalent, and that they are not used to industrial life in general. However, the village people themselves believe that the real reason is that the villagers are very well organised, and that the employers fear and try to avoid potential labour trouble. Hence the fairly large numbers of migrants presently working in the industrial sector of New Bombay.[14]

The data of the *Village SES-96* reveal that the large majority of village families do have some kind of income, but in nearly 70% of households there is only a single earner. Over one in four of all males (26.6%) earns less than Rs. 1,250 per month, and another 47.2% earns between Rs. 1,251 and Rs. 2,650 per month. Thus, if nearly 70% of households have a single earner, and nearly 74% has a monthly income of at most Rs. 2,650, most village households must be having a monthly family budget of maximum Rs. 2,650. Moreover, the *Village SES-96* gives an 'excellent' example of an error that is frequently being made in questionnaire surveys. According to the SES data, 53.4% of all males did 'not mention' their income, and of all the females, even 96.1% did 'not mention' their income. Since in the entire *Village SES-96*, a category 'not applicable' has nowhere been included, it is almost certain that all those women who do not have jobs and were asked for their income, have been included in the category 'not mentioned'. This means, in other words, that of the 53.4% of males who did 'not mention' their income, a large part is probably not just refusing to answer but is simply unemployed and not having an income. Therefore, the real figure of unemployed males in the villages of New Bombay is probably much higher than the figures of the SES suggest. Anyhow, the fact that nearly three in four (73%) of all village households have 4 to 7 members, and that nearly 70% of all households have just one earner, is in itself indication enough of the high rate of unemployment.

The unemployment rate seems to be especially high among the younger generation. In all of the villages that were visited, the number of young males (< 30) that are employed in a permanent job is estimated at just 15 to 20% of that age group. Older PAPs who earlier managed to get an alternative job after land acquisition are in many instances still permanently employed, but since the village population has since those days increased by nearly three times (mainly due to natural increase and partly to marriages and migration), the pool of predominantly young job seekers has increased manifold. The increase in the number of jobs, on the other hand, has been much slower, and the combination of neglect of the authorities and reluctance of industry has resulted in a situation in which the majority of the village population (and the large majority of youngsters) is either unemployed, underemployed, temporary employed, or self-employed.[15]

Whereas the situation has been fairly difficult for most PAP-households, for those families that lost their land to the MIDC or the JNPT, and for those who did not have any land (and were thus not considered as PAPs), the situation has been even worse. Land-holding village households in MIDC or JNPT areas received the limited compensation money, whereas landless households did not get any compensation nor benefits. Yet, most of these people have also lost their jobs and incomes, and have in many instances been much more affected than the formal PAPs.[16] It is not only the poor agricultural labourers and tribals who have been in this situation. The fairly large group of fishermen in the area has also suffered from these developments. Due to the sharp increase in chemical pollution in the TBIA zone since the 1960s, the fish population in the coastal areas has almost disappeared. To enter deeper waters, more advanced equipment is needed and special permits are required, both of which are too expensive for most fishermen. In the southern area of New Bombay the same phenomenon of decreasing fish population has been observed, only here mainly due to the construction of huge walls in the new harbour. The fish which used to come into New Bombay waters with high tide has disappeared. Consequently, in the whole of New Bombay, fishing as a profession has almost been wiped out. Some fishing is still being done, but only as a life-saving activity, until a proper job can be found. Moreover, the quality of the fish has become very doubtful and consumption of it may be dangerous.

Cost and quality of living   Thus, whereas a large number of families in the villages have lost their lands, their jobs and their incomes, very little alternative employment has been generated. In the meantime, the general level of inflation in the area has sharply increased, and each and every commodity and service now has a price. In earlier days most families were active in agriculture, in fishing, or in both, and food for personal consumption was therefore largely available for free. Even five years ago, rice that would have cost Rs. 5 per kg. is now costing Rs. 15 per kg, although the quality of the rice has become much worse. Earlier, many households also had cattle and thus milk for free, but since there isn't any grass left and thus no possibility to keep cattle, the price of milk has gone up from Rs. 1.75 per litre ten years ago to about Rs. 11 at present.[17] Fruits also used to be available almost for free as the area was covered with mango trees. Now, fruit has become a very expensive commodity. Firewood, which was earlier used for cooking, has now been replaced by expensive kerosene from the market. And finally, to quote a fisherman, 'earlier there was a lot of good fish at a small price, now there is little bad fish at a high price'. The direct result of this rapid inflation in food products is that the diet of many households has drastically changed, with less fruit, less milk and less fish on the menu.[18]

Apart from such food related price increases, modernisation and urbanisation have also caused a rapid inflation at several other levels: the sharp increase in the cost of education has been mentioned before, the expenses for medical treatment have been rising, and the cost of basic services such as water and electricity has become many times higher since the NTDA started to provide these services. Earlier, water used to come from the village well, was almost free of cost, and in normal circumstances available around the clock. Electricity and a sewerage system were often provided by the 'Panchayat' (local council), sometimes with the help of the 'Zila Parishad' (District authorities), for which a small amount of money (Rs. 50 to 60, according to the size of someone's plot) was charged on an annual basis. Later, from the 1970s, CIDCO introduced a charging system for water and electricity according to the meter, and the cost of electricity increased from around Rs. 0.40 per unit meter ten years ago to Rs. 1.50 per unit now. Within the same period, the cost of water increased from around Rs. 25 per two months to about Rs. 150.

Overall, the large majority of households in the villages that were visited admitted that, financially, life has become much more difficult now than it used to be. And things seem to be going from bad to worse and becoming unbearable to many. Recently, a property tax was introduced by the newly established NMMC, which has been causing both anger and fear among the people. This property tax, which was to become the major source of income for the new corporation, has been fixed at an extraordinary high 23% of the estimated value of the house for residential property and 35% for commercial property.[19] Obviously, as few people can afford to pay such large sums of money on an annual basis, a lot of people refuse to pay. What will happen now is still not clear, but the NMMC does have the power to evict people from their houses.

Beside the overall increase in the cost of living, the fall in environmental standards, with pollution levels sometimes reaching dangerous levels, has added to the overall decreasing quality of life for the village population. The poisoning gas emissions of the chemical industries in the TBIA, and the emissions of the thousands of cars and trucks plying the major road running next to it, right through the residential heart of New Bombay, have become so severe that even the local and state authorities now admit that there is a serious problem. It has been estimated that because of industrial and traffic pollution the general temperature in the area has gone up by about 8 degrees compared to ten years ago. In combination with the very poor and ever worsening sanitary conditions in the area, this has given rise to a serious malaria problem. A number of interviews with local doctors has revealed that malaria is the major reason for people to visit a doctor's surgery.[20] Next come all sorts of pollution-related problems such as pneumonia, bronchitis, burning eyes, and even digestion problems. Children and elderly people are the most vulnerable groups. It is estimated by the local people and by a number of village doctors that life expectancy in New Bombay has come down from about 80 years ten years ago, to about 60 years now; figures which to some extent are being supported by regionwise birth and death rates in the state of Maharashtra. Further, as has been mentioned before, the waters in and around New Bombay have become so polluted by all kinds of chemicals that hardly any fish is left, and that the fish that is being caught in the nets is often found dead or sick, and smells of chemicals.[21]

To conclude, it is not surprising that almost all villagers consider the overall standard of living to be much lower now compared to what it was 20 years ago. Although most families are presently living in a more spacious brick house instead of the small mud houses they used to live in,[22] the cost of living has gone up sharply, and the quality of living and of the environment has drastically gone down. In the meantime, in many households the small amount of compensation money has been spent, and no alternative jobs have come their way, so that the overall condition in which many families presently find themselves is far from optimistic.

Village rehabilitation and physical integration   The Development Plan stated that the original villages of New Bombay were physically not to be adversely affected by the development of the nodes, but were instead to be provided with a minimum of physical and social infrastructure, and were to be fully integrated into the new urban environment.

One of the most striking visible features of growth and development of New Bombay, is the physical impact of the growing nodes on the villages. Today, after 25 years of project implementation, a number of nodes are growing and expanding so rapidly that they directly threaten to swallow up the nearby villages (photograph 7.1). At several locations, the multi-storied buildings of a node are now within the village periphery. The photograph below gives an indication of the phenomenon, and shows the physical expansion of the new CBD and its threatening impact on one of the villages. Moreover, within the villages that are located in the vicinity of an expanding node, land or property is gradually being taken over by outsiders and real estate developers, and in several villages two and three-storied buildings are mushrooming and sometimes replacing the old village houses (foreground of photograph). Thus, internally the villages are also rapidly changing. These developments will be discussed in the next section.

Photograph 7.1: One of the Belapur villages physically threatened by expansion of New Bombay's CBD

Such developments are of course not only taking place in New Bombay. Every present-day large city has been physically growing and expanding, thereby incorporating and integrating former agricultural land and villages. However, the nature and the consequences of this process differ from city to city, and former developments in existing cities may be an indication of what can be expected in New Bombay in the near future. Jain (1994) for example, writing on the 'urban villages' in Delhi's periphery, has observed similar patterns and mechanisms than the ones that are currently in operation in New Bombay. In the Union Territory of Delhi, so far 111 out of a total of 357 villages have been incorporated into the urban limits, these 111 having a population of about 600,000 and covering a total area of about 1,500 ha. Another 53 villages are likely to be covered under the extended urban limits of Delhi by the year 2001. In the case of Delhi, however, the authorities have never explicitly claimed to protect these peripheral villages from the effect of increasing spatial expansion of the capital, as the NTDA did in New Bombay.

Not surprisingly Jain also observed that in the urban villages of Delhi 'the speculative forces and interests have generated large-scale commercial and industrial activities in these settlements' (Jain, 1994:180-81). Furthermore, he added that 'public amenities are supposed to be provided by the development authorities when a village is urbanised.' These however often take years, and 'even if provided, are too meagre to the demand' (Ibid:181). As a result, Jain concludes, 'In absence of any curbs or controls and without any sort of infrastructure to support such urban functions, the environment deteriorates' (Ibid). In short, earlier developments in the rural periphery of Delhi thus seem to have followed a similar path as in New Bombay today.

Above we saw that in terms of social rehabilitation there was a clear distinction between the population of the villages that were located on CIDCO land, and those that were located on MIDC and JNPT land. The same distinction applied to the rehabilitation of the villages internally. In addition, the package of social and physical infrastructure that was to be provided also differed according to whether the villages had physically been replaced or not.

Nava Sheva (or New Sheva) is an example of a village in the southern area of New Bombay that was completely replaced from its earlier location due to the development of the JNPT port. When this shift took place 10 years ago, most basic urban utilities such as running water, electricity, a basic sewerage, and toilet blocks were provided by the JNPT.

The maintenance became the responsibility of the local Panchayat. A school, a small dispensary and a temple were also provided, but by the Raigad District authorities. In this case, although the population in this village is having the same problems of unemployment as other village people, internal village rehabilitation seems to have been fairly reasonable. The latest example of a village that is to be relocated soon is Kukshet gaothan (in MIDC area close to Nerul node), officially for safety reasons.[23]

Nevertheless, the number of villages that have been, or will be shifted from their original location is limited. The large majority of villages in New Bombay have not been physically replaced, but have been affected by the growth of New Bombay in an indirect way. Although the proposal was that, where needed, they would be provided with basic physical and social infrastructure, the NTDA's performance in these villages has been much less convincing and far insufficient.

In terms of physical infrastructure and basic utilities, far too little has been provided. Whereas electricity has been available in most villages for quite some time (mostly provided by the Panchayat), the majority of villages still have to do without piped water. In some villages water has been provided by the Panchayat since the 1960s or was taken from the village well or lake. In other villages water has indeed been provided by the NTDA, but is being supplied only two hours a day, or is limited to just one central distribution point at the edge of the village (such as in Pagote) from where water has to be collected.[24] Similarly, the villages where a properly working and well-maintained sewerage system is in operation are few; either such a system is completely absent (in the majority of villages), or is in a very poor state and not properly working. Moreover, in contrast to the sewerage systems in the nodes, which are half closed, the gutters in the villages are open and therefore dangerous, and are a major contributor to the generally poor sanitary conditions in the villages. Further, toilet blocks and urinals may indeed have been provided in a third to half of all the villages (as the NTDA claims, see table 7.2), but are poorly or not maintained, and are just another source of hygienic problems.[25] Finally, approach roads have, according to the NTDA, been provided in about a quarter of the villages, but in very few villages is there any public transport provision. Most villages are not serviced by bus, and the nearest bus stop or train station may be a fifteen to twenty minute walk away.

In terms of social infrastructure, things are not much better. Most of the money that CIDCO has spent on village rehabilitation has gone to school buildings, which are predominantly used for primary education, which in itself is of course a positive thing.[26] Very little however has been spent on other social facilities. Primary health centres for example have presumably been considered as absolutely unimportant, since they have been provided in only one village, notwithstanding the fact that in most villages minimum 25% of the children is suffering from malaria. Playgrounds have been provided in just two villages, a library and a gymnasium in just one village each, and community centres in one out of ten villages (table 7.2).

One could say, therefore, that the overall performance of the NTDA in village rehabilitation has up and till now been quite poor, and way removed from the extensive rehabilitation package that the affected villages were promised. If the objective to physically integrate the villages into the new city is a genuine one, providing a few badly maintained toilets or a central water distribution point is not going to be sufficient to serve the purpose. As the situation is now, the general impression of the local people is that the NTDA is not really concerned about what happens to the villages and to themselves, and that the planning authority simply has other priorities. How the system of village rehabilitation is working in practice, is explained by one village Sarpanj :

> If we want certain facilities for our village such as a temple, a community centre, or a sewerage system, we have to require those things from CIDCO Bhavan. But down there it takes years before a file passes all necessary desks and has all the necessary stamps. Why? Because at regular intervals the file will get stuck, and if that happens we know someone is expecting money before the file will be passed on. We have become used to this, and in such cases the village pays whatever is required, as we do not have much choice.

Similar (spontaneous) complaints of widespread corruption could be heard in most villages and from many people, but the extent to which such practices are in operation is obviously difficult to say. In general, when speaking to the local people about the NTDA and its rehabilitation programmes, feelings of disappointment and betrayal prevail.

The Gaothan Expansion Scheme (GES) and the 12.5% scheme Beside the rehabilitation packages that focused on the socio-economic integration of the village population and on the physical integration of the villages, a third kind of rehabilitation had to do with the ownership of land. Earlier in this chapter we discussed the introduction of the first land recovery regulations in 1978, when the first GES was introduced. Under this scheme PAPs could get 5% of their earlier acquired land back. Eight years later, in 1986, the GES scheme was amended and former landowners could get 12.5% of their land back, first at FSI 0.75 (as it still is in the village areas outside the scheme), later at FSI 1.5 with 15% allowed for commercial use. The price to be paid to get this land back was twice the acquisition cost plus Rs. 5 per m² as land development charge (for roads, street lighting, water supply, etc.). It is evident, that similar to most other rehabilitation measures discussed earlier, this 12.5% scheme was only applicable to former landholders, and thus excluded all those categories of the population that earlier did not posses any land. Moreover, no single category (landholders included) within the MIDC and JNPT controlled areas was entitled to the scheme, as it only applied to the CIDCO controlled areas.[27] The official nod from the state government to start releasing lands under this new scheme finally came, after long delays, by late 1994. Under this scheme, 59,752 people would get 12.5% of their land back. In March 1996, about 5,000 of these had received their plot of land. Transfer of these lands by the allottees to third parties was legally restricted.

How successful has this new scheme been in terms of rehabilitation of the local people, and to what extent are they profiting from it? In the northern developed half of New Bombay, implementation of the 12.5% scheme is well under way, but in the southern half of the area it has yet to start.[28] However, no matter in what area and in which village, for most people this scheme has come far too late to be helpful. Only those who had a big chunk of land before and/or are still having enough money can really benefit from the scheme. Most people, especially those who did not find an alternative job afterwards, have been using the compensation money to survive. Now the NTDA requires twice the earlier cost of acquisition, plus an additional cost for land development per square metre. What does this mean in practice? It means that if a landowner has lost a big plot of 1 ha to the NTDA in the 1970s, he probably received about Rs. 37,000 as compensation (at the normal compensation rate of Rs 15,000 per acre). Today, if he hasn't been able to find another job, that money will have been spent on food, education, medical expenses, etc. Nevertheless, he would be entitled to get back a plot of 875 m², for which he would have to pay Rs 37,000 x 2= Rs. 74,000 + Rs. 4,375 (as land development charge) = Rs. 78,375 in total.

Unfortunately this is not all. Three more elements are equally important. First, in most cases the former landholder would have to share this new plot of land with a number of family members, say 10 to 20, which means that each person's plot will be reduced to 40-80 m² maximum. Secondly, the 12.5% which is always being put forward would in practice never be more than 8.75% net, as about 4% of that land is being used for infrastructural development. Thirdly, and most importantly, land recovery regulations and building regulations have become very strict. Those who get land back under this scheme will have to get a building permission within six months and construction has to be fully completed within four years. If this can not be done, a heavy delay charge of 25 to 40% of the total amount will have to be paid to the NTDA.[29] Moreover, new building regulations determine that a house built on such plots may not be constructed using wood for instance (which would cost about Rs. 50,000), but has to be of cement (which for the same size would cost about Rs. 600,000). Thus, the kind of money that is needed under this scheme to get land on the one hand, and to construct a house on it on the other far exceeds most village household budgets, even for those who used to have fairly big plots of land and/or are having a regular income from employment.

Naturally, what happens is that private developers approach the villagers with offers for an alternative choice, even in such far-off and undeveloped regions as Uran. They mostly propose to take over adjacent plots of several households on a 50-50% basis, in which the plot owners get a flat 'with all modern facilities' in a new multi-storied building as well as one of the commercial spaces on the ground floor. The other flats and possible additional commercial space are being sold off to third parties.[30] Many households who cannot afford to follow any other path ultimately give in to these proposals, gaining some money and a possible income from their small commercial space, but losing the land. They either move in to this new building themselves and try to make a living from the small commercial space, or rent out this commercial space and/or flat and stay in their original village house.[31] The restriction on transfer of the lands by the allottee encourages such unauthorised transactions.

Some villagers have followed still another path. They have simply sold their new plot and have taken the money in hand. Some households have even sold everything, land and village house, and have moved out of the village and/or the New Bombay area.[32] According to a group of older fishermen in Vashi, at least 75% of all the land that was reserved for the village people under the 12.5% scheme, has already been transferred to some project developers. Nevertheless, the large majority of local people still live in Vashi gaothan, as they do not sell their village houses.

Only a small minority has managed to finance the entire thing by themselves. They are the families that earlier had a big plot of land, have managed to find an alternative job after land acquisition, have strong family ties in which every family member contributes to the cost of buying the plot and building the house, and/or have been clever enough to come up with a solution to get the money without putting everything at risk.[33]

At present, a number of sites in the villages, or in-between the villages and the nodes, have been outlined for implementation of the 12.5% scheme. Twenty-one sites have been reserved for this purpose in the Thana-Belapur area, one in the Uran area (Dronagiri), and seven in the Panvel area. At the time when the scheme was started, most beneficiaries could get the plot of land they wanted, but afterwards a stiff competition for the commercially most valuable locations has set in, and few people manage to get the land they want. Some may get land within their own village at a commercially interesting spot, others may get land miles away from their village and in a commercially undeveloped area.

A permission of the NTDA is still required to start exploiting the 15% commercial space of the newly acquired plot, which in practice frequently proves to be a major hurdle. According to a former village Sarpanj this is, again, mainly because of the widespread corruption within the NTDA which, according to this former village head, 'is currently at its peak'.[34] Apparently, CIDCO officers seem to be deliberately delaying approval to start exploiting the commercial spaces until the new owners come up with sufficient money. If a number of owners who have joined hands start building shutters in their new building without permission the construction will be torn down. Consequently, as with the village integration scheme, people do not have any choice but to pay whatever money that is required. Obviously, for some households this additional financial burden makes the scheme even more impossible than it perhaps already was.[35]

Thus, whereas at present the scheme is only profitable to a very limited number of former landholding households, a large number of households have not yet received any land. Moreover, thousands of families (non-landholders in CIDCO area, and all people in the MIDC and JNPT areas such as Kukshet or Karalpada gaothans) were not included in the scheme anyway, and have not benefited from anything. Only recently a small-scale scheme has been introduced, in which affected non-landholding families would also get a small plot of land (40 m²). According to the

NTDA, as per March 1996, 42 out of around 1,000 households had benefited from this scheme.

These different land recovery schemes, and the commercial dynamics which have resulted from them, obviously have a very large impact on the villages in terms of physical outlook. In many villages, such as Vashi gaothan, old houses are being removed and replaced by new houses having two or three stores.[36] Elsewhere, and mostly on the outskirts of the villages, adjacent plots of land under the 12.5% scheme have been taken over from the new landowners by a private developer and a new large residential-cum-commercial building is being constructed. These buildings, although they are being built on village land, have the same characteristics of similar buildings in the nodes. Consequently, due to these developments the physical difference between nodes and villages is becoming increasingly vague, and the villages are gradually being physically integrated into the new city; not because of CIDCO's village rehabilitation and integration schemes unfortunately, but because of simple commercial mechanisms.

Conclusion

With the start of development of the New Bombay area, the large majority of local people lost their land, their jobs and their incomes. Only the former landholders in the area administered by CIDCO were given a modest financial compensation at the then market rate. On the other hand, as Banerjee-Guha already in the late 1980s observed, over the years very little alternative employment has been provided, not by the authorities or public companies, nor by the private industries in the area (1989:183-84). Consequently, unemployment and underemployment, and the lack of a proper and steady income, are the major problems for the original village population. In the meantime, the cost of living (for food, education, transportation, medical expenses, and all kinds of taxes and services) has increased many times, and most people have spent their compensation money on meeting these expenses.

Beside employment, most other rehabilitation schemes which had initially been introduced were soon abandoned, and the major part of CIDCO's expenses on rehabilitation has been put into school building construction, instead of rehabilitation schemes from which the entire population would directly benefit. Moreover, not a lot more has been done within the villages, as most villages are still having grave problems with such basic utilities as piped water or a properly working sewerage system, causing all kinds of sanitary and health problems. The GES and 12.5% scheme are good initiatives in themselves, but are coming far too late for the majority of village people, as they do not have the necessary finances to buy the plot of land they are entitled to, and/or to build a house on it within the time limit and according to the strict (and expensive) building regulations. Consequently, a large-scale buying-out process is in full swing, with the villagers selling their plot in the private market, gaining some money, but ultimately not really moving forward; a similar process in other words as has been observed in many other large cities in India, with that difference however that the NTDA in New Bombay had explicitly planned to protect the villages in New Bombay from the physically adverse effects of urbanisation.

For a further evaluation on rehabilitation and integration of the former villages and their population I refer to chapter 9. In the following chapter we will first have a look at the major developments, mainly socio-economically, that have occurred in the nodes of New Bombay in the course of the last twenty-five years.

Notes

[1] How stiff the opposition against acquisition sometimes was, and how violent the police reaction, became evident in January 1984 when a demonstration of villagers of Pagote went completely out of hand and five people were shot.

[2] As development in the northern part of New Bombay is almost fully completed, the impact of urban development on the villages in this part of the region is obviously very different from the impact on the villages in the southern part of New Bombay, where development is still in its infancy.

[3] *Transect walks*: a fieldwork technique based on systematically walking with informants through an area, observing, asking, listening, discussing, identifying different zones, seeking problems, solutions and opportunities, etc. (Chambers, 1992:16-17).

[4] My initial intention was to have one male and one female translator. Unfortunately, no female candidates applied for the job. Nevertheless, my doubts of, and fear for a possible female reluctance in the villages was not justified at all.

During the day, when many men were gone to work, most village women were quite happy to sit with us and respond to our questions.

[5] An excellent introduction on how to construct an argument on the basis of qualitative data is being provided by Mason (1997) chapters 6 and 7, and by Coffey and Atkinson (1996), pp. 54-80/139-162.

[6] The three most commonly used forms of qualitative data organisation are indexing (cross-sectional and categorical), non-cross-sectional data organisation, and diagrams and charts. Analysing and explaining qualitative data, on the other hand, needs to be done, as Mason has put it, 'with rigour, with care, and with a great deal of intellectual and strategic thinking' (1997:162). It must be noted, however, that until recently, almost all of the published literature on qualitative research focused on methods for generating data, and although there are now some very useful contributions about how you might analyse such data and construct explanations and theories on their basis, the territory is still rather sparsely explored.

[7] One ha. or 10,000 m² equals 2.471 acres => compensation of Rs 15,000/acre = Rs. 37,065/ha or Rs. 3.70/m².

[8] The compensation rates that are mentioned here, are the rates that were paid to the majority of people. It has however to be noted that some people have received significantly less, whereas others managed to receive significantly more, even when dealing with the same authority. The rates that were paid have thus not been entirely uniform for all the villages and all the people. Why this is so is not exactly clear, but has in part to do with the exact location of the plot (road side or not), but may also have to do with the extent of opposition, or the personal status of a PAP.

[9] The stamp-duty (a tax on certain legal documents) depends on the amount of the case. This has largely been the dominant factor in deciding whether or not to take a case to court, and when it has been taken to court, and in deciding how much the requested compensation would be.

[10] The legal system is such that when the head of a landholding household dies, it is his elder son who legally gets the right to the land, and in New Bombay it is therefore also the elder son who gets the possible benefits provided by any rehabilitation schemes. Hence the numerous court cases within families.

[11] It was primarily D'Souza, the then MD of CIDCO, who was responsible for the rehabilitation of the PAPs and who intervened on the villagers' behalf.

[12] Going from Sanpada village to nearby Vashi Modern College by bus for instance will cost a student Rs. 3.50 one way, or some Rs. 1,750 per year. Moreover, some schools and colleges in the nodes are located such, that taking a rickshaw is almost a necessity. In such case the transportation cost more than trebles.

[13] The NTDA's Chief Rehabilitation Officer however claims that the village youth is 'a very difficult group to employ', because they all want to get white collar jobs and are, according to her, 'very selective in their professional choice'.

[14] An independent research paper on migration into New Bombay was prepared by Prasad and Gupta in 1995.

[15] Alcoholism is a very serious problem in the villages of New Bombay. It would be interesting to investigate to what extent this is a result of the unemployment situation and the overall low standard of living.

[16] According to CIDCO's Chief Rehabilitation Officer, children of affected people who did not possess land can since recently also benefit from training programmes.

[17] From an interview with the last cattle holder and milk merchant of Turbhe gaothan.

[18] Although no one was speaking for himself, numerous people seemed to be convinced that many villagers who earlier had two square meals per day, do no longer have this certainty.

[19] In practice this means that a village household whose house is valued at a modest Rs. 50,000, has to pay not less than Rs. 11,500 to the local authorities every year.

[20] One village doctor noted that the large influx of construction workers from the state of Rajasthan is an additional factor in the spread of malaria in the area.

[21] 'Earlier the water used to be as blue as this net, now it is black' (quote of a fisherman in Vashi gaothan).

[22] According to the SES-95, at present 60.2% of all village houses is *pucca* housing (bricks), and another 35.2% is semi-pucca housing.

[23] Although Kukshet village has been there for hundreds of years and was joined by a chemical plant right next to it in 1965, there was never any talk of shifting the village. Now suddenly the village has to be relocated 'for safety reasons'.

[24] 'If the water now stops, there is no water. Before we had the well'. (quote of a 72-year old former land cultivator from Nava Sheva)

[25] In some villages such as Kukshet, the collection point of the sewage system of several nodes ends up dangerously close to the village.

[26] In many villages a primary school had already been established by the Panchayat and/or District authorities before the NTDA moved into the area. Therefore, the role of the NTDA was sometimes limited to financing a decent school building or adding a number of rooms.

[27] During a recent visit of Chief Minister Manohar Joshi to New Bombay on 20 April 1997, several more promises were made, one of which is that the 12.5% scheme would be extended also to the landowners in the area controlled by the JNPT (*The Indian Express*, New Bombay supplement, 3 May 1997).

[28] According to P.C. Patil, leader of the PAP organisation, 80 to 90% of the scheme has been completed in Turbhe, Vashi, Sanpada and Shiravane gaothans, about 50% in Koparkhairane and Belapur gaothans, and about 25% in Diwe, Airoli, Nerul and Salsole gaothans.

[29] According to CIDCO's Lands and Surveys Officer, in the fifth year the delay charge will be 25% of the total amount, in the sixth year it will be 35%, and in all further years 40%.

---

[30] The proposals with which the builders' agents approach the villagers are outlined in great detail, with the exact size of the flat, the exact interior design, the use of the materials, etc. One of the reasons why people do not easily accept such offers is because they fear a defaulted construction or a poor quality of the building.

[31] The major condition for this alternative private scheme to work is that the plot has to be significantly large and will therefore usually need the approval of several households.

[32] Especially in Vashi gaothan (predominantly a fishing village) this seems to have happened on a certain scale, and many people just seem to have no idea of what the current rates for the land are. Some households seem to accept an amount of money which is way below the market rates (Rs. 75,000 or even less). Further south, in Sanpada, a family was offered over Rs. 600,000 for their newly acquired 100 m² plot of land.

[33] One man for instance, who got back a plot of 750 m², sold off the commercial space on the ground floor at Rs. 2,000/ft² before construction had even started.

[34] This is not the same critical sarpanj who accused CIDCO officers of corruption in relation to village rehabilitation.

[35] As a result, people who have joined hands now choose to start constructing the building without the shutters, as to give the impression that the building is not going to be used for commercial purposes.

[36] According to the *Village SES-96*, nearly half (48.8%) of all the houses in the villages have been renovated at an average cost of Rs. 40,766, and another 22.3% of the houses are newly built at an average cost of Rs. 57,988.

8 New Bombay, a City for All?

'The effort has been to avoid the spectacular, to provide minimally for the affluent few and to promote the convenience of the greatest numbers. New Bombay, then, will not be another grand city, it will be a city where the common man would like to live'

(From the foreword of CIDCO's first chairman in the NBDP)

Introduction

At the time when the New Bombay concept was taking shape, it was quite clear that the new twin city was planned to become a city in which every citizen, rich or poor, would be able to live. One of the main objectives of the project put forward by the state government stipulated that the new town was 'to reduce disparities in the amenities available to different sections of the population'. Some time later, the words that were used in the foreword of the NBDP also pointed towards a town that would be planned in such a way that it would benefit 'the common man', and not just a small elitist section of the population.[1] The NBDP stipulated that 'every family living in New Bombay shall have a dwelling of its own, however small', and that New Bombay 'will not be another grand city, it will be a city where the common man would like to live' (CIDCO 1973:18).

The central question in this chapter is not whether 'the common man' would *like* to live in the new town of New Bombay, but rather to what extent he *can* live in this new town. In a first part of this chapter we shall therefore draw a socio-economic profile of the New Bombay population, and in a second part we will proceed with a basic explanation of the socio-economic dynamics that have been at work in the area. Such a socio-economic profile would uncover to what extent New Bombay's current population corresponds with 'the man in the street'. In 1995, over 59% of the public housing stock of about 95,000 houses (i.e. 81% of the total housing stock) had been built by the NTDA to accommodate lower income groups. Furthermore, in every node several residential sectors were to be reserved almost exclusively for low-income families, and a large sites and services programme was to be developed to provide shelter exclusively to

New Bombay's urban poorest: the so-called Economic Weaker Sections. Given these general planning and housing objectives, the expectation would be that the majority of households currently living in the new town would belong to the lower income categories (EWS and LIG, see further).

In order to draw such a socio-economic profile of New Bombay's urban population, we have been using a large data set, which was collected for the NTDA in 1995-96. It was based on a large socio-economic questionnaire survey (hereafter called 'SES-95/96') which had a random sample of 25% of all households living in the nodes of New Bombay, i.e. a total of 19,727 households numbering 79,075 persons in total.[2] The people who took the questionnaires for the NTDA were properly trained and instructed, and the data processing was done by an independent agency. Although the data indicated a few errors in the questionnaire design as well as in the data processing phase, the overall quality of the survey was quite good and the results are trustworthy. The data that were collected have been tabulated to two different levels: to the general level (all nodes together), and to the nodal level (data per node). Further, data were also collected in New Bombay's sites and services areas.

Four data series have been selected for analysis in this chapter :
(1) the general data (for all the nodes);
(2) the data for one specific (randomly selected) node, i.c. CBD Belapur
(3) the data for three so-called 'low-income sectors', which were specifically planned to accommodate low-income groups;
(4) the data that were collected in the 'sites and services' of New Bombay, which were outlined and reserved for the lowest income group (EWS).

For the nodes in general and for Belapur node, the data series that were processed for the NTDA have been used; for the three low-income sectors the raw data have been re-processed; and for the sites and services the data of the SES-95/96 have been used as well as the data of a personal interview survey in these areas. In 1987, eight years before the SES-95/96, a similar survey had been conducted and some comparisons over time can thus be made. The list of the 32 standardised tables that have been produced from the raw questionnaire data has been included as an appendix.

Socio-economic profile of the population in the nodes

The nodes in general

Socio-cultural characteristics   Of the total number of 79,075 people included in the sample, 53.95% were males and 46.05% were females. The largest group fell within the category of 25 to 44 years, followed by the 10 to 15 year olds. Over 53% of the nearly 20,000 households included in the sample count 4 to 5 persons, over 31% 2 to 3 persons, and another 12% 6 to 7 persons. The large majority of the population (85.5%) consider themselves as Hindu, 4.8% as Muslim and as many as Christian, and 2.8% as Sikh. The mother tongue is predominantly Marathi (48.7%), followed by Hindi (14.2%), Malayalam (8.0%), Punjabi (5.0%), Gujarati (4.5%), Tamil (3.7%), and Kannada (3.1%). Thus, only half of the population in the nodes of New Bombay speak the local Marathi language.

Housing characteristics   In 1995, over 81% of the total housing stock in New Bombay was built by CIDCO, of which the major part (59%) is between 16 and 35 m² in size. Of all the houses that have been built by the private sector (nearly 19% of the housing stock), almost 68% is between 51 and 100 m². Nearly 32% of the total housing stock (CIDCO and private) was rented from the private market, nearly 31% was purchased outright from CIDCO and 9% from the private market, nearly 15% was purchased under the hire-purchase formula, and 7.5% has been resold.[3] The average amount of money that was originally paid for a house purchased from CIDCO was Rs. 137,000, and from the private market Rs. 231,000. Of the 2,875 households that bought a house under the hire-purchase formula, 18.5% of purchases were financed by personal savings, 25.2% by a loan from HUDCO, and another 20.7% by a combination of both. Of all the households that took a loan to purchase the house, the average monthly income per household was Rs. 4,779. Of the 9,272 households that bought a house under a different scheme (outright purchase from CIDCO, from the private market, resale), a large part (nearly 40%) was financed by personal savings, and another 14% by savings in combination with a loan from the employer. Nearly 26% of the current housing stock was occupied between 1986 and 1990, and a quarter of the stock was occupied in 1994 alone.

Economic characteristics   Of the total sample of 79,075 persons, 26,233 persons (88% of which are males) are employed and are having an income.

The main labour categories among males are the manufacturing industries (29%) and petty shops or businesses (13%), and among females, education (20%) and the manufacturing industries (11%). The income levels of male and female workers/employees (table 8.1) show that at least 91.5% of males and over 70% of females have an individual income of at least Rs. 1,251 per month. Over 67% of males and nearly 48% of females have an income of at least Rs. 2,651 per month, and nearly 31% of males and over 17% of females have an individual income of at least Rs. 4,451 per month.

Table 8.1: Monthly income of workers/employees in New Bombay (nodes)

Income category	Males	Females
Rs. 1250 or less	5.8%	17.1%
Rs. 1251 - 2650	26.3%	28.6%
Rs. 2651 - 4450	34.3%	24.7%
Rs. 4451 - 7500	23.7%	14.3%
Rs. 7501 - 10,000	6.0%	2.3%
Rs. 10,001 - 15,000	0.9%	0.4%
Rs. 15,001 and more	0.3%	0.2%
Not mentioned	2.7%	12.4%

Based on data SES-95/96

Figure 8.1: Income of male and female workers/employees in New Bombay nodes

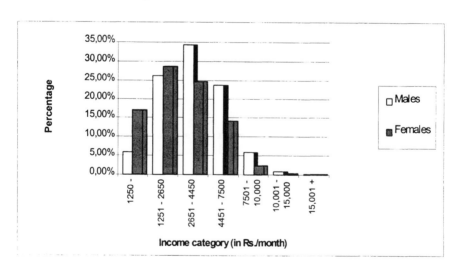

Over 74.3% of all households (14,427) have a single earner, 19.5% have two earners (3,803), 4.5% have three earners (882), and 1.5% have four or more earners. Every household living in the nodes is thus having at least one earner in the family. However, it must be noted that 1.54% of households did 'Not Mention', which means that given the fact that a separate category 'Not Applicable' was not included, part of this small percentage may have been unemployed, as they could not be put into any other category. Of the 53% of households that have 4 to 5 persons, the average combined income per household is Rs. 3,507. Of the 31% of households that have 2 to 3 members, the average combined income per household is Rs. 2,133. Of the 12% of households that have 6 to 7 persons, the combined household income is Rs. 5,584 on average. The overall average income per household per month for the entire population in the nodes is Rs. 4,932.[4] If a distinction is being made between the households living in CIDCO housing and those in private housing, figure 8.2 shows that the income levels of families living in private housing are markedly higher than the income levels of families living in CIDCO housing.

Figure 8.2: Household income per category in public and private housing

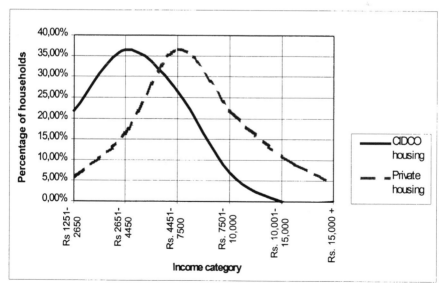

Table 8.2: Household income per category in public and private housing

Income category	CIDCO housing (16,093 hhs.)	Size (av.)	Private housing (3,634 hhs.)	Size (av.)
upto Rs 1250	2.1 %	26 m²	0.8 %	71 m²
Rs 1251-2650	21.9 %	23 m²	5.5 %	41 m²
Rs 2651-4450	36.3 %	29 m²	16.2 %	63 m²
Rs 4451-7500	26.5 %	38 m²	36.6 %	69 m²
Rs 7501-10,000	6.9 %	48 m²	22.0 %	74 m²
Rs 10,001-15,000			11.0 %	82 m²
Rs 15,000 +			4.7 %	87 m²

Based on data SES-95/96

That the average income per household in all the nodes of New Bombay is nearly Rs. 5,000 per month, and that every family in the nodes has at least one regular income (with about three quarters of families having one income and one quarter having at least two incomes) is quite revealing. Furthermore, figures on education, another significant socio-economic indicator, show that 37.4% of the total population in the nodes has finished (or is studying up to) at least 10th standard, and that nearly 10% is having a graduate or postgraduate diploma or equivalent, or is studying for one (table 8.3). Finally, it is also interesting to note that of the 19,727 households included in the sample, just 41 households (or 0.21%) are Project Affected People (PAPs) who were originally living in the villages.[5]

Table 8.3: Level of education (total population nodes)

Pre-primary	12.4 %
Primary	28.1 %
Middle (up to 7th standard)	21.3 %
Secondary (10th standard)	19.8 %
Higher secondary	7.5 %
Graduate (or studying for graduate)	5.5 %
Studying engineering, architecture, medicine, law	2.4 %
Diploma engineering, nursing, architecture	1.0 %
Postgraduate or studying for postgrad.	0.8 %

Based on data SES-95/96

CBD Belapur node

The general perception among people in New Bombay is that some nodes have gained the status of 'upper class node' (with a large part of the population belonging to the upper social classes), and some others the status of 'low class node'. Vashi is an example belonging to the first category, Airoli an example belonging to the second category.[6] CBD Belapur, the node which was randomly selected for analysis, probably falls somewhere in-between. It has been developed later than Vashi, and was supposed to become the new CBD of New Bombay. However, as we have seen in chapter 6, this new CBD is in reality not more than a conglomerate of government and semi-government offices with very little private business activity. Is the socio-economic profile of the population in this node different from the general pattern?

The number of households in CBD Belapur included in the sample in 1995 was 2,034 of which 53.6% were males and 46.4% were females. Almost 80% of the population was Hindu, 6.7% Sikh, 5.6% Christian, and 4.0% Muslim. Marathi was the most spoken language (40.8%), followed by Hindi (16.5%), Punjabi (9.6%), Malayalam (8.3%), and Bengali and Gujarati (both 4.0%). The population of CBD Belapur is, similar to the entire New Bombay nodal population, quite young: over 38% is between 25 and 44 years of age, nearly 14% between 10 and 15, and another 11.4% between 16 and 21. Only 13% is older than 45. Over half of all households (54.2%) count 4 to 5 members, 30.7% count 2 to 3 members, and 12.5% count 6 to 7 members. Exactly two families living in CBD Belapur (or 0.1% of the sample) are classified as PAPs.

In terms of housing, nearly 91% of all the houses in Belapur node are built by the NTDA. About half of the housing stock has been occupied before 1990, the other half after 1990. Nearly one in five of all houses (19.5%) has been bought from the NTDA under the hire-purchase formula for an average cost of Rs. 25,000, and another 8% has been purchased outright for average cost of Rs. 153,000. Exactly 7% has been purchased outright from the private market at an average cost of Rs. 374,000. Remarkable in Belapur node however is that 22.2% of houses has been resold, and that 37.7% of houses is being rented. Unfortunately, no resale prices have been mentioned. Over 33% of rented houses was rented out for less than Rs. 500 per month, nearly 34% for Rs. 500 to 1,000, nearly 29% for Rs. 1,000 to 1,500, and 3.8% for Rs. 2,500 to 5,000 per month, although these figures do not say much in practice.[7]

Nearly 71% of all households in Belapur node have a single earner, and nearly 23% have two earners. Here as well, 3.8% of households did 'Not Mention', which in part may point to unemployment. Of all working males, the majority (59.3%) is earning between Rs 2,651 and Rs. 7,500 (29.5% Rs. 2,651-4,450, and 29.8% Rs. 4,451-7,500). Over 8% is earning more than Rs. 7,500 per month, and just 5.4% is earning Rs. 1,250 or less. The income levels of working females are in general significantly lower. Not surprisingly, the main occupation of males is in Belapur node in government offices and other public sector offices (34.7%), in manufacturing (16.2%), and in private offices or in a personal business or shop (7.5 and 8.3% resp.). Women are also mainly employed in government offices and other public sector offices (20.3%), in education (15.6%), and in private offices (9.5%). Another 9.2% of females is working in domestic service.[8] Making the distinction in income levels between those households living in CIDCO housing and those living in private housing, a fairly different pattern emerges when compared to the overall New Bombay pattern. In CBD Belapur there does not seem to be as much difference between income levels, and households staying in CIDCO housing even seem to have a slightly higher overall income than those staying in private housing (figure 8.3). This may, on the one hand, be due to the fact that a large proportion of those families staying in private housing are only renting the house, and on the other hand to the fact that many of the government employees in this node have been accommodated in CIDCO housing.[9]

Table 8.4: Household income per category in public and private housing in Belapur node

Income category	CIDCO housing	Private housing
Rs 1251-2650	12.9 %	8.1 %
Rs 2651-4450	26.1 %	32.4 %
Rs. 4451-7500	34.3 %	33.0 %
Rs. 7501-10,000	11.8 %	16.2 %
Rs. 10,001-15,000	5.2 %	7.0 %
Rs. 15,000 +	2.3%	0.0 %

Figure 8.3: Household income per category in public and private housing in Belapur node

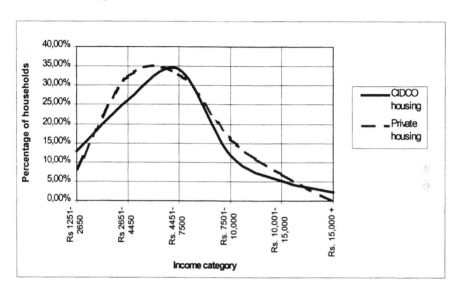

Household-wise, the average income per month is Rs. 5,787, which is 17.3% higher than for all the nodes together (see further). This generally higher income level in Belapur node is reflected in the ownership of private vehicles, which is higher than in New Bombay in general. Whereas for all the nodes together 11% of all households owns a scooter and 6.5% a car or a jeep, in Belapur node these figures are 18% and nearly 10% respectively. So far as education is concerned, the number of people that have received a fairly high education is in Belapur node also slightly higher than in the nodes in general.

In short, the socio-economic profile of the population living in CBD Belapur node is not significantly different from the general profile of the entire New Bombay population: in socio-cultural terms the only difference is the relatively large group of Punjabi and Sikh people residing in Belapur; in terms of housing, the most striking difference is the high percentage of dwellings that has been resold and the high percentage of houses that has been rented out; in terms of income levels, wages are significantly higher in Belapur node and even slightly higher among those staying in CIDCO housing; and the level of education is also slightly higher than the New Bombay level in general.

Significance of the data

To understand the impact of these figures, for the nodes in general as well as for Belapur node, we first have to look at the definition of different income categories which are used by HUDCO. It is on this basis that the NTDA has defined income categories in New Bombay. The table below shows what has, over the years, been understood by the terms Economic Weaker Sections (EWS), Low Income Group (LIG), Middle Income Group (MIG), and High Income Group (HIG).

Table 8.5: Income groups in New Bombay

Applicable since...	EWS	LIG	MIG	HIG
January '86	up to Rs 700	Rs 701-1500	Rs 1501-2500	Rs 2501 +
April '91	up to Rs 1050	Rs 1051-2200	Rs 2201-3700	Rs 3701 +
April '93	up to Rs 1250	Rs 1251-2650	Rs 2651-4450	Rs 4451 +

Thus, according to the latest revision, a household in New Bombay which is currently having a monthly income of Rs. 3,700 for example belongs to the MIG. Now, if the figures on household incomes (all nodes) are brought into these categories, it will be observed that almost no households (all together just 359 out of 19,207 households) belong to the EWS category (figure 8.4).[10] The number of households belonging to the LIG category is higher (all together 3,727 out of 19,207 households), but still less than one in five (19.4%) of all households, and the number of households belonging to the MIG category is with 33.4% still higher (6,423 out of 19,207). Most significant however is that the largest group of households (45.3%) belong to the HIG category (8,698 out of 19,207). In other words, the distribution of households in New Bombay is such that almost four in five households (78.7%) belong to the higher income groups (MIG+HIG), and only 21.3% to the lower income groups (EWS+LIG).[11]

Figure 8.4: Population of New Bombay nodes by income group (1995)

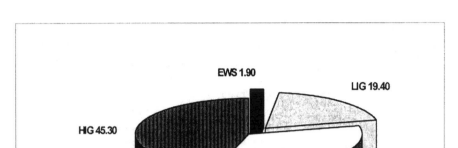

In 1987, when the first Socio-Economic Survey was conducted, the percentage of households in New Bombay belonging to the higher income groups was with 54% already relatively high, but far less than what it is today (figure 8.5). The average monthly household income in 1987 was Rs. 2,112, or less than half of what it is today (Rs. 4,932).

Figure 8.5: Population of nodes by EWS/LIG - MIG/HIG category, 1987+1995

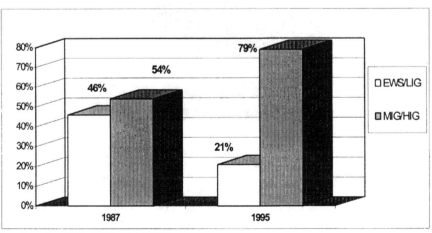

Based on data SES-1987 and 1995/96

Comparing these data with similar figures in South Bombay, which is generally considered the richest part of Greater Bombay (and presumably one of the most affluent regions in the whole of India) is quite revealing (figure 8.6). These figures, which give the household income distribution for South Bombay in 1989 (more recent figures were not available), have therefore been linked to the income categories of EWS, LIG, MIG and HIG as they were between 1986 and 1991 (see table 8.5 above). In this way, it can be observed that in 1989, 9.4% of households in South Bombay had a monthly income of less than Rs. 750, 33.7% had an income of Rs. 751-1,500, 26.1% had an income of 1,501-2,500, and 30.8% had an income of Rs. 2,501 or more. Thus, in other words, in 1989, 43.1% of all households in South Bombay were belonging to the lower income groups (EWS/LIG), and 56.9% of households were belonging to the higher income groups (MIG/HIG).[12] For the entire area of Greater Bombay (a separate figure for South Bombay was not mentioned), the average household income in 1989 was Rs. 2,468.

**Figure 8.6: Population of South Bombay belonging to different income
categories (1989)**

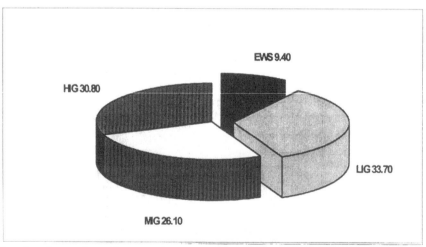

Based on figures *Socio-Economic Review of Greater Bombay, 1993-94*

The only figures on household income distribution that have been found for urban India at large, are based on completely different categories and can therefore not be compared. It is nonetheless worth noting that based on 1988 and 1989 figures, 71.9% of all urban households in the country had a monthly income of maximum Rs. 2,083 (Statistical Outline 1994-95). Thus, only the upper section of this 71.9% (having an income between Rs. 1,501 and Rs. 2,083) would *not* belong to the lower income groups (EWS/LIG).

In short, household income distribution figures in New Bombay in 1987 were fairly similar to the same figures in affluent South Bombay in 1989. Since then, the percentage of households in New Bombay belonging to the higher income groups has sharply increased to nearly 79% (figure 8.5). What in the meantime the evolution in South Bombay has been is not known. Furthermore, the overall average household income figure for New Bombay in 1987 (Rs. 2,112) was, given the two-year difference in data collection, quite comparable to that of Greater Bombay in 1989 (Rs. 2,468). Quite remarkably, this 1989 figure for Greater Bombay is almost exactly half the average household income in New Bombay in 1995 (Rs. 4,932).

In general it can be concluded that the population currently living in the nodes of New Bombay is a fairly affluent population. It is argued by some (among whom several NTDA officers) that the income categories which are used by HUDCO and CIDCO are simply unrealistically low and that they therefore give a distorted picture of the supposed affluence of New Bombay's population. However, the figures for South Bombay in 1989 (as well as indirectly those for urban India in 1988-89) refute these arguments. Moreover, sometimes it is also claimed that the relatively large proportion of high income households in New Bombay, compared to South Bombay and urban India as a whole, is a result of the higher percentage of formal sector employment in New Bombay. This argument is not convincing either: whereas the percentage of the informal economy in Greater Bombay is generally estimated at 60-70%, in New Bombay it is estimated at not less than 50-60%.[13]

Some low-income sectors within the nodes

CIDCO has been providing eight different types of housing, from type A to type H, all having specific characteristics such as size, approximate price, eligible monthly income, registration account, and design characteristics. Apart from the houses falling under sites and services schemes, house types A, B and C (up to 34 m² in size) have been built for the lower income groups, whereas types D, E, F, G and H (between 40 and 100 m² in size) have been built for the higher income groups.[14]

In addition to the nodes in general and to CBD Belapur node, we have also been looking at the socio-economic profile of the people currently living in housing type A, B or C in a number of selected low-income sectors. For this purpose, the raw SES-95/96 survey data were re-processed specifically for these sectors. In Vashi, the first developed node, we have focused on sector 2, in Nerul on sector 10, and in CBD Belapur on sector 2. All three sectors were planned as to accommodate households belonging to the EWS and LIG categories.[15]

Sector 2 in Vashi Of the 637 houses in Vashi's sector two, 285 were part of New Bombay's first small-scale sites and services scheme and were thus planned for EWS households, whereas the other 352 houses (160 B-type, 96 C1-type, and 96 C2-type houses) were planned for LIG households. Figure 8.7 however shows that in reality only 8.6% of households currently living in this sector fall within the EWS/LIG income categories for whom these houses were planned, and over 91% of households in this sector actually belong to the higher income groups (MIG/HIG). No single household falls into the EWS category, and not less than 67.3% of all households in this sector has a monthly income of more than Rs. 4,451 and thus falls into the HIG category. In this sector, exactly 1/3rd of the houses has *officially* been resold, and nearly half (48.3%) of all households have moved into this sector coming from Greater Bombay (55.2% of whom from South Bombay).

This upper socio-economic profile is indirectly reflected by education, by ownership of vehicles, and by occupation categories. Education-wise, 60.3% of the people in this sector went to school up to at least 10th standard, and the number of people that have (or are currently studying for) a diploma of higher education is 15.5%.[16] In terms of vehicle ownership, over 8% of households owns a car, and another 11.5% a scooter. Occupation-wise, about 94% of all males are employed. Over 30% of all working males are involved in business or working as a professional, and 20.2% are skilled and unskilled workers. Nearly 79% of females are not working outdoors.[17]

Figure 8.7: Household incomes in sector 2 in Vashi

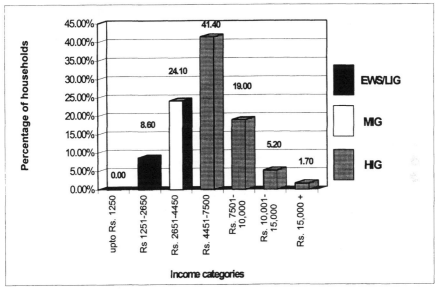

Source: based on SES-95/96 data for Vashi node (based on Report No. 21)

To avoid any distortion resulting from the fact that sector 2 of Vashi, due to its location right opposite New Bombay's main commercial sector, is an extreme and exceptional case, two other low-income sectors were selected. That sector 2 of Vashi is an extreme case is indeed correct. Nowhere else in New Bombay is a low-income sector located so close to a major commercial centre. Nevertheless, the figures for sector 10 in Nerul node and for sector 2 in Belapur node prove that although the figures are less extreme, the same phenomenon of relatively few low-income households can be observed.

Sector 10 in Nerul In sector 10 in Nerul, 99.3% of the 706 houses that were included in the sample is public housing built by the NTDA. Over 34% of these houses have been bought under the NTDA's hire-purchase formula, nearly 30% have been bought from the NTDA outright, and nearly 33% are rented. Of these 706 houses in the sample, about 81% were planned to provide shelter to EWS or LIG households, whereas the remaining 19% were meant for higher income households (MIG and HIG).[18]

In terms of current household incomes, however, the figures in this sector are not radically different from what they are in Vashi's sector 2. Just 36.5% of households living in sector 10 can be categorised as lower income households (EWS/LIG), and 63.5% actually belongs to the higher income groups (MIG/HIG). Moreover, although the plan had reserved less than 2% of the houses for the HIG category, in reality not less than one in four houses (25.2%) is being occupied by an HIG household. The figure below shows the discrepancy in Nerul's sector 10 between what was planned and what the current situation is.

Figure 8.8: Sector 10 in Nerul: planned and real occupation of houses based on household income

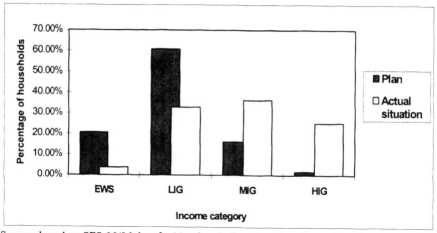

Source: based on SES-95/96 data for Nerul node (based on data Report No. 21)

Sector 2 in Belapur   This sector, in which houses were constructed to accommodate exclusively families having an income of maximum Rs. 2,650 (falling into the LIG category), is another example that shows the gap between plan and actual situation. Of the 79 houses included in the sector-sample, 40 were *officially* resold and 12 were rented out. Presently, just one in four of the households living in sector 2 of Belapur (25.7%) belong to the LIG category for which they were built, and no less than 40% of the households living here actually belongs to the HIG category.

   In short, in all residential sectors of New Bombay where housing has been specially provided and reserved to accommodate low-income families, the majority of houses is in practice occupied by middle income and high income households.

Sites and services in Airoli node

The Bombay Urban Development Project (BUDP) is an extensive programme, co-financed by the World Bank, in which sites and services are provided to the lowest income groups of New Bombay. When the project is completed, over 22,000 of such sites and services must have been constructed, spread out over six nodes of New Bombay. In a first phase of the programme (BUDP-I), 3,500 plots were provided in Airoli node, in a second phase (BUDP-II) another 10,800 were provided in Koparkhairane node (photograph 8.1), and in the third and last phase (BUDP-III) close to 8,000 are planned in New Panvel, Nerul, Kalamboli and Kharghar nodes.

Photograph 8.1: Sites and services in Koparkhairane

Under this programme, EWS households were given the opportunity to buy a serviced and subsidised plot of land at RP of 25 to 50%, depending on the size of the plot. Loans could be taken from HUDCO at relatively low interest rates (5-12%). The initial contribution was not to exceed 1.5 to 2 times the average monthly household income, and the monthly instalments afterwards were not to exceed 15% of that average. Different types of plots (SS-1, SS-2 and SS-3) were provided, each measuring 30 m². SS-1 is just a plot with a plinth of cement, SS-2 is a plinth with walls, and SS-3 a plinth with walls and a roof (photograph 8.2). In some areas such as sector 3 in Airoli, a slightly different formula has been used under the name AR-1 (photograph 8.2 next page). In this alternative form, 15 m² of the 30 m² plot has been built-up by the NTDA with a small and very basic house, and owners are allowed to use the other 15 m² according to their wishes. On all SS and AR-1 plots, facilities for a toilet have been provided, aiming to make the sanitary situation in these BUDP areas better than what it is in the slums of Bombay. All plots are serviced with water, electricity, and with a sewerage system, and the entire area has paved roads and street lighting.

One of the major features and advantages of this scheme is that construction materials and time of construction are basically free, giving low-income families the opportunity to build how and when their financial situation allows it. Allotment of the plots by the NTDA was (and is) done by a booking system (lottery), for which potential buyers have to register and a computer makes the final draw. The conditions to be eligible for such sites and services plots are: (a) a permanent residence in the state of Maharashtra for the last 15 years; (b) having a household income that does not exceed Rs. 1,250; and (c) showing proof of formal employment.

The SES-95/96 in sectors 2, 3 and 4 in Airoli   For our analysis of the socio-economic profile of the population in these BUDP areas, we have selected the SES-95/96 data for the three sectors in Airoli where sites and services have been provided, i.e. sectors 2, 3 and 4. They were selected because they were part of the first phase of the BUDP project (which took off in 1986), and are therefore the oldest and most developed part of the programme.[19] These three sectors were to cater almost exclusively to EWS households. The SS-1,2,3 and AR-1 houses in the interior of the sectors (nearly 97% of houses measuring 30 m² maximum) were constructed for the EWS only, whereas housing on the major road side was meant for higher income groups (the overall financing system being based on the principle of cross-subsidisation).

Photograph 8.2: AR-1 plot in Airoli

Of the 906 households that were included in the sample and were living in SS-1,2,3 or AR-1 houses, 44.6% (4,057 people) admitted that they are not the first owners of the house, and another 30.6% were just renting the house. The average *official* resale price was Rs. 60,871, which as we mentioned earlier is a manipulated figure that does not reflect the real situation. Most significantly, just 3% of households presently occupying these houses belong to the EWS category, and another 27.2% to the LIG category. Thus, in total, less than one in three houses in these sectors (30.2%) is currently occupied by a lower income household, and just below 70% of the houses is occupied by a higher income family (MIG 41.4% and HIG 28.4%). A detailed overview of household income levels in the sites and services of Airoli is included as an appendix.

**Table 8.6: Household income in sites and services in sectors 2, 3 and 4
of Airoli node**

Household income		Percentage of households
Rs up to 1250	(EWS)	3.0%
Rs 1251-2650	(LIG)	27.2%
Rs 2651-4450	(MIG)	41.4%
Rs. 4451-7500	(HIG)	23.8%
Rs. 7501-10,000	(")	3.6%
Rs. 10,001-15,000	(")	0.7%
Rs. 15,001 +	(")	0.3%

Based on data SES-95/96. Note: of the 906 households just 21 (or 2.3%) did not
mention the combined income

Figure 8.9: Household income distribution in sites and services in Airoli

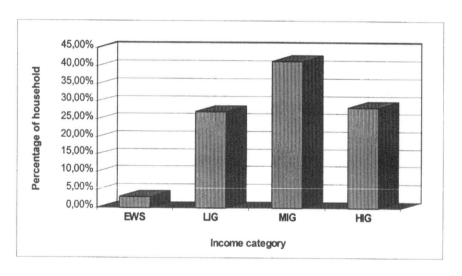

The people currently occupying these low-income houses have
mainly come from Airoli itself (29.4%), from the Eastern Suburbs of
Bombay (20.3%), and from villages in New Bombay (11.8%). Their main
motivation to shift to the sites and services of Airoli was the easy
availability of accommodation on ownership (41.5%), termination of an
earlier housing agreement (10.2%), and easy availability of rental
accommodation (8.9%). Over 71% of households in these sites and services
have a single earner, while another 22.7% have two earners.

The interview survey in sectors 2, 3 and 4 in Airoli As the questionnaires were not anynomous, and as some of these SES-95/96 data for the sites and services were not very accurate and sometimes difficult to interpret,[20] an additional survey based on interviewing and transect walks was conducted in these three sectors.[21] The perception of the local people and of the real estate agents in these three sectors reflect to a large extent the data of the SES-95/96. A very large majority of people currently living in sectors 2, 3 and 4 believe that at least 70% of the first owners (those families who originally lived on the plot) have vacated their house and have moved out of the area.[22] Two main reasons are given for the fact that the large majority of first owners have vacated their house: the first reason is financial difficulties, the second pressures from outside.

The general perception is that about one-third of those households that have moved out did so shortly after they were allotted a plot by the NTDA. Although initially they had the opportunity to buy a house at subsidised rates and to take a cheap loan, mostly that loan did not exceed Rs. 3,000, which is only a fraction of the money that was needed to construct a house and still be able to pay for all other expenses. In 1986, Rs. 40,000 was considered an absolute minimum to finance the cost of construction of a house on a SS plot that would be fit to live in.[23]

The other two-thirds of low-income households that have moved out did so after some time, usually when construction of the house was finished. In such cases, the main reasons were also financially motivated, but sometimes also socially and professionally (lack of space for expanding families, transportation problem to and from work due to the poor public transport links between Airoli and the job centres of Bombay's Eastern Suburbs). Many families eligible for an SS or AR-type plot bought such a plot from the NTDA, took the maximum loan from HUDCO, and took additional loans from a private source or used personal savings to pay for the cost of construction. However, these families were also confronted with high additional expenses for urban utilities: water and electricity supply, service charges, and more recently the property tax.[24] Consequently, numerous families had difficulties to repay their loans and to pay for all other expenses at the same time. Many have lost their property to the borrower, whereas others gave in to the pressures of estate agents, or sold their house to a third party, for which they received a much higher rate than the one for which it was originally bought.[25]

To conclude, New Bombay is, thus, far from being *the city for all* that it was supposed to be. At present, the new town is predominantly catering to the middle and high income groups, and over time, lower income families, even in those areas that were specially reserved for such households, have increasingly been removed by higher income families. The main question then is why the number of low-income families in New Bombay is decreasing? What are the main factors that contribute to the process of rapid socio-economic change?

Factors responsible for the high percentage of high-income groups

In an attempt to explain why New Bombay has become a city predominantly catering to higher income groups, increasingly pushing out low-income households, four major factors can be distinguished: (1) the NTDA's land and housing policies and its attitude in relation to the working of the land and housing markets; (2) the NTDA's policies in the provision, management and maintenance of infrastructure and services; (3) the overall cost of living in New Bombay; and (4) the homogeneous economic base of the new town.

The NTDA's land and housing policies and the working of the land and housing market

The NTDA's policies in housing construction   In this chapter only the formal housing sector will be discussed. Informal housing (slums and squatters) has not yet become an inseperable part of urban life in New Bombay, as is the case in most Indian towns and cities. An official estimate has put the number of people living in informal housing in the new town at 60,000. This figure is relatively modest compared to many other Indian cities. The reason why this is so is in part because of the NTDA's 'cleanliness policy'. The new town is nicely planned, developed and maintained, and the NTDA is doing whatever is necessary to keep it that way, i.e. frequently using its eviction power to keep the city 'shack-free'.

The effect of the NTDA's privatisation drive since the late 1980s had a fairly strong impact on housing construction. Up until 1987, CIDCO had followed the conventional approach to housing construction in India. The corporation constructed and maintained the public housing stock largely by itself, and involvement of the private sector was limited. Since 1987, however, the private sector has increasingly been involved in housing construction.

Two large housing schemes, the Demand Registration Scheme (DRS) and the Participatory Developers' Scheme (PDS), were a first expression of the new housing construction policies. The DRS was launched first, in order to assess the demand for shelter in New Bombay, and to plan for meeting that demand. Over 55,000 applications for a house were received and construction of 19,500 dwellings was started. This was a much bigger housing construction scheme than had ever been undertaken before, and the NTDA therefore decided not to construct these houses by itself (as it used to do) but to engage private sector consultants (architects, developers, etc.) and construction firms. They were to implement twenty housing schemes of about 1,000 to 1,200 tenements, for a prescribed fee, and at a cost of Rs. 100 to 150 million each. Contracts were to be awarded to the highest bidder, and were to be supervised by eminent project management consultants instead of CIDCO's in-house staff. Some time later the PDS was launched. In this scheme reputed builders and developers were pre-qualified and each of them was given 3 ha. of developed land. Nearly 35% of that land was to be used for providing houses as per the NTDA's specifications, at a pre-determined rate to be paid by the NTDA, and to be marketed by it. The other 65%, of which 5% was allowed to be used for commercial purposes, could be used by the developer to construct houses according to his wishes.

Since these schemes, the private sector has been a major participant in public housing construction in New Bombay, while the role of the NTDA has been limited to facilitating, co-ordinating and overall supervision of the projects. In general, however, this privatisation drive in housing construction itself does not seem to have had major negative effects for the public, as the total cost of these schemes was fixed in advance. Much more important in this respect is the impact of the methods that have been used by the NTDA in selling land and housing to the public, and its laissez-faire attitude vis-à-vis the developments in the land and housing markets.

The NTDA's policies in selling land and housing   Prior to 1987, plots and houses in New Bombay were sold by the NTDA at fixed price. From the moment the market picked up, however, the NTDA's methods in selling land and housing have drastically changed. To provide housing for businesses and for high income groups, the NTDA now puts developed plots of land on the market which are being sold to the highest bidder, resulting in high-price housing catering exclusively to the higher income groups. On the other hand, the NTDA is also allotting plots to housing societies formed by specific target groups, resulting in what it claims to be 'affordable housing for all'. Whereas in the first case the land is being sold by auction, in the second case it is being sold not at fixed price, but by tender, at a minimum percentage of the reserve price. If a plot does not fetch that minimum percentage, the plot will not be sold.

The system used by the NTDA, if planned in a balanced way and if properly executed and controlled, could work, and could indeed provide housing to all. In reality however, these land sales methods have caused a spurt in land and property rates, the more since land and housing have been released in small bits to keep the prices in the market as high as possible. Consequently, the situation at present is that even the cheapest houses have become so expensive that they are far beyond reach of the lower income groups; a process which was already observed in the late 1980s by Banerjee-Guha (1989). Officially, the NTDA argues that a general increase in land and housing prices, and the sale of commercial and high-income residential space by auction, helps them in collecting the necessary funds to subsidise low-income housing. However, the number of plots and of housing that is effectively being sold by tender to low-income groups is, given the proportion of these groups, fairly small. Moreover, getting hold of a CIDCO house has increasingly become a matter of luck, as the allotment of such houses is literally done by random computer draw (and therefore open to possible corruption and forging).[26]

The most important consequence however of the sharply increasing land and property rates is that the subsidised public housing stock in New Bombay has become a major target in the private market, and that the NTDA in practice does not do anything to prevent a rapid selling-off process of the public housing stock.

Working of the land and housing markets and the NTDA's laissez-faire attitude   Since the early days of the New Bombay project, the NTDA had introduced certain rules and regulations to prevent a selling-off process of public housing within a rapidly urbanising environment. The sales contracts of both 'outright purchase' and 'hire-purchase'[27] (the two systems by which the NTDA sells public housing) stipulate that public property cannot simply be resold to a third party. The outright purchase agreement says in article 12 that

> The purchaser shall not without the previous permission in writing of the Corporation let, sub-let, sell, transfer ... The Corporation may grant such

permission to the purchaser subject to such terms and conditions as may be specified ... including the condition for payment of additional price ...

The hire-purchase agreement, on the other hand, says in article 21 that

> Until the Deed of Apartment is executed under the said Act in favour of the Hire-Purchaser, the Hire-Purchaser shall not sell, transfer, assign or part with his interest under or benefit of this agreement in any manner in favour of any person or persons.

In other words, someone buying a house outright from the NTDA cannot sell it unless transfer charges are paid to CIDCO, and someone buying a house under the hire-purchase formula cannot sell it unless the full price of the house has been paid. The amount of such transfer charges depends on the size of the house and on the location: For sector 10 in Nerul for instance these charges go from Rs. 6,720 (for a house of up to 20 m²) to Rs. 44,320 (for a house above 70 m²) .

In the case of resale of a CIDCO house, the official transfer charges are not the only cost involved. In addition, the stamp-duty must be paid, and a property tax and service charges will have to be paid (both of which are based on the official value of the property).[28] Therefore, in order to reduce these expenses, the real sales figures are usually manipulated. Thus, in short, a lot of housing transfers are being done illegally: either under the table (mostly in case of hire-purchase), or officially but with manipulation of the sales figures (often in case of outright purchase).[29] Such illegal or semi-legal operations have in New Bombay occurred on a large scale, and have led to a serious discrepancy between the real housing situation and the situation as it is according to the official documents. It is here that the data (in the SES-95/96 as well as other data) are not reliable.[30]

According to figures of the NTDA of September 1994, nearly half (48.9%) of the then public housing stock (i.e. just over 81% of the total housing stock) was built for the EWS/LIG income category, 36.3% for the MIG category, and just 14.8% for the HIG category. Figure 8.10 shows the official figures (SES-95/96) of public housing occupation by different income categories. According to these figures the real proportion of EWS/LIG households living in CIDCO housing is just 24%, or less than half of what was planned; the proportion of MIG households is according to these figures 36.30%, which is right on target; and the proportion of HIG households living in CIDCO housing is thought to be 37.30%, or about two and a half times larger than what was planned.

Figure 8.10: New Bombay population in CIDCO housing and private housing belonging to different income groups

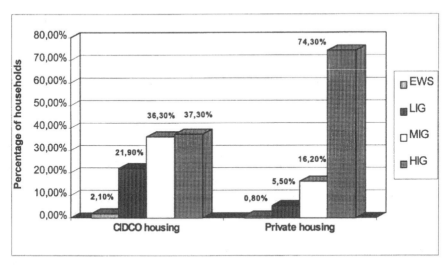

These official figures, based on survey data, thus already show a significant gap between public housing occupation as it was planned and occupation as it is in reality. However, due to illegal sales of such public housing on a large scale, the real situation is far worse than these figures indicate.

To find out the extent of first owner occupancy in public housing in New Bombay, and to check what the current prices for housing in the market really are, a small additional interview survey was done with real estate agents and households in a number of selected sectors.[31] Some interesting results have come out of this. First, it has become clear that in recent years (up to 1996) the impact of the private market on the value of public housing in New Bombay has been dramatic. Secondly, it has become clear that the impact of the NTDA's laissez-faire attitude in the housing market has been equally dramatic, with low-income households increasingly being replaced by high-income groups.

The impact of the private market on the value of public housing is amply illustrated by the price evolution of houses in a number of residential sectors in the nodes of New Bombay. If we go back to the three low-income sectors that were discussed earlier, very sharp price increases can be observed, and a much more serious displacement process than is officially thought becomes apparent. First, the small houses on the 30 m² plots in Vashi's sector 2, which in the mid-1980s were sold by the NTDA for Rs. 8,000, were in February 1996 in the private market sold for Rs. 700,000 to Rs. 1.3 million.[32] Whereas the official resale figure (SES-95/96) has been put at 22%, real estate agents and first occupants in this sector estimate that the number of original households is not more than 15 to 20%. About three in four of the original houses in this sector have been removed by the new occupants and replaced by new, often two-storey dwellings. About six years ago the NTDA doubled the FSI in this sector and has formally allowed transfer of these houses. Thus, displacement of both households and dwellings have, so far as this sector is concerned, been done legally (see also chapter 9).

Secondly, in the more recently developed sector 10 of Nerul, the number of first owners is estimated to be around 15%. Unlike sector 2 in Vashi, houses here have not been modified physically, although prices have followed a similar path. An A-type house which was initially Rs. 16,000 was sold in 1990 for Rs. 30,000 to 45,000 and was by 1996 costing Rs. 180,000. Prices of E1-type houses (built on the outside of sector 10 and meant for LIG or lower MIG) have gone up from Rs. 90,000 in 1990 to Rs. 260,000-300,000 (Feb 1996). The initial price of the largest type of houses in sector 10 (F type, measuring 100 m²) is not known, but their price in February 1996 was around Rs. 1.5 million.

Thirdly, in sector 2 of Belapur node, where 12 rows of 26 A-1 type houses (measuring 18 m² and meant for LIG) were constructed in 1978, the initial price of a house was Rs. 8,000. In 1992 however the NTDA allowed an increase in FSI, and the price in the market immediately shot up to Rs. 190,000. By February 1996 these very small houses were being sold for about Rs. 450,000. This actually means that the price of these houses has become higher than the price of a B-10 type of house, the price of which increased about three-fold compared to its price of Rs. 100,000 in 1992. For the entire sector 2 it is estimated that about 70% of the A-1 type houses has been resold.

Such sharp price increases have not been restricted to low-income sectors only. In sector 3 of Belapur for instance, which constitutes a mix of houses, from B-10 type (upper LIG) to F type (HIG), local real estate agents estimate that about 50% of the houses are still owned by the first owners. The general selling pattern is such, that the lower the income category is, the stronger the selling process, and vice versa.[33]

Further, in the sites and services areas that were discussed earlier, the same sharp increases in property rates have occurred, and a similar displacement process of low-income households could be observed. Plots in the three sectors of the BUDP project in Airoli which ten years ago were sold by the NTDA for Rs. 8,000-16,000 (depending on the type of the house and the time of selling), were already in 1990 resold in the private market for about Rs. 35,000 to 45,000 (SS1), Rs. 45,000 to 60,000 (SS2), and Rs. 60,000 to 80,000 (SS3). Six years later, in January 1996, these plots/houses were sold for Rs. 125,000-150,000 (SS1), Rs. 160,000-190,000 (SS2), and Rs. 200,000-270,000 (SS3).[34] For those plots which are located in or close to the local markets, prices go up to Rs. 350,000 to 400,000. Initially, the plots in these sectors were exclusively for residential purposes and no commercial activities were allowed, adjustments to the houses were not allowed and a full second floor could thus not be built, and such plots could not be sold.

Notwithstanding the fact that these sectors were frequently visited by officers of the NTDA to check for illegal extensions or for commercial activities, many houses were enlarged and hundreds of small businesses were established on these plots.[35] Moreover, in these areas also the same illegal resale procedures as in the nodes have been in operation, and the strict rules that were to prevent a resale of plots and houses in the sites and services have never seriously been enforced. This displacement process is not a recent phenomenon, as Banerjee-Guha (1991) already observed in the mid-1980s (when the first sites and services plots were allotted). Later, the NTDA largely abandoned its own rules concerning a restriction on resales in these areas, bringing the legal situation in fact merely in line with the practical situation. With this move the NTDA admitted its failure to protect even these sites and services from the market forces, and to secure a minimum of shelter for the lowest income groups in New Bombay.

Finally, not only the resale figures have experienced such exorbitant increases, rental rates have also gone up significantly. One example: in 1988 a C-5 type house could in Belapur be rented for about Rs. 650-700. Four years later the rate had almost doubled to Rs. 1,200, and another four years later (by February 1996) it had further increased to Rs. 2,000. A so-called RH-4 (row house) was as recently as 1991 rented out for Rs. 1,500, whereas in early 1996 the price was Rs. 4,000 on average.[36]

By comparison, table 8.7 gives the annual inflation figures for India as a whole between 1992 and 1996 for both Consumer Price Index (CPI) and Wholesale Price Index (WPI). It demonstrates that the general inflation level over the last four years has been relatively moderate, and only a fraction of price increases in rental and property rates.

Table 8.7: Annual inflation in India 1992-1996

	CPI	WPI
1992-93	6.1%	7.0%
1993-94	9.0%	9.3%
1994-95	8.5%	9.0%
1995-96	9.7%	5.0%

Source: http://www.prakash.org/economy/statistics

To conclude, the displacement of low-income households in New Bombay, even in those areas which were specifically planned for the urban poor, has been (and is) very significant, and a large proportion of property transfers has been done illegally. Chiefly responsible for this development are the policies of the NTDA, which has never seriously tried to prevent such illegal practices (which obviously push up market rates). Quite the contrary, as of late sales regulations and limitations have gradually been revised and have been made more flexible, and selling property is no longer a problem, with the obvious result that the number of low-income households in New Bombay will further decrease. The question is, what the use of a plan is if the policies that have been adopted are characterised by such a laissez-faire attitude?

The NTDA's policies in the provision, management and maintenance of infrastructure

Beside the NTDA's land and housing policies and its attitude in relation to the land and housing markets, the agency's policies in the provision, management and maintenance of infrastructure and services have been equally important to explain the recent process of rapid socio-economic change in New Bombay. Earlier, the NTDA was planning, developing, and maintaining New Bombay's infrastructure and services. Later, however, its role has increasingly been limited to planning and co-ordination, whereas development and maintenance have increasingly been privatised. The areas which have been prone to 'contracting out' are numerous, and include general environmental sanitation jobs such as road sweeping, cleaning of shopping complexes and markets, maintenance of the railway stations, garbage collection and waste management, and the maintenance of sewage treatment plants; further also the collection of CIDCO's service charges (by CBOs such as senior citizen's clubs or 'Mahila Mandals'), the maintenance of the water supply system, the development and maintenance of parks and gardens, the maintenance of street lighting, the management of educational and public health facilities and, indirectly, part of the public transportation sector.

This privatisation has had certain positive effects for both the NTDA and the public. For the NTDA it has mainly given positive financial and operational effects, as no staff had to be employed on a permanent basis, and as potential labour trouble with heavy delays was avoided.[37] On the other hand, for the local population it has given some positive effects in terms of small work contracts for local businesses with guaranteed minimum income levels for the workers (*some*, because in practice usually only experienced agencies are qualified), and it has resulted in more efficiency in service delivery and in community participation, without an increase in the cost of these services.

Although this public-private partnership in New Bombay between the NTDA on the one hand, and the private sector and the public on the other, has given certain advantages, its overall effect for the public is not entirely positive. No reasonable arguments can be put forward against the privatisation of urban infrastructure and services in countries or regions with limited public resources, *on the one condition* that the mechanism is to the benefit of the general public. Privatisation is clearly not a universal policy option that can be applied at any level, at any time and at any place. Its impact has to be thoroughly researched in advance, its consequences clearly understood, and its price effect kept under strict control before it can be implemented. If key urban and social services like education, public health and transportation are partly or fully privatised, generating major negative cost effects to the public (as has been the case in New Bombay), such privatisation may become unacceptable.

In the initial stages of development of New Bombay, the NTDA constructed school and hospital buildings by itself and had some control on the activities within these buildings. Later however its role was limited to simply providing a plot of land, or at best constructing the building, whereas managing the school or hospital was left to a private trust. However, as there are no municipal schools left in the nodes, all privatised schools (so-called kindergartens included) are now (unofficially) charging high donations (sometimes called 'money for the building fund') of Rs. 2,000 to over Rs. 10,000, mostly on top of the annual tuition fees. In contrast, Bombay counts numerous good municipal schools which are basically free of cost.[38]

The same evolution could be observed for hospitals and dispensaries, most of which are now being privatised. Although New Bombay has a few municipal hospitals, they are so poorly equipped that if people cannot afford a private hospital, they have to go to one of the municipal hospitals in Bombay for treatment. The cost of medical treatment in New Bombay in such privatised hospitals or dispensaries, which are sometimes called *charitable* hospitals where the overall quality of treatment is quite good, has over the years seriously increased and has become unaffordable to many.[39]

Finally, CIDCO's claimed 'exciting privatisation experience' has also (indirectly) caused major expenses for the public in transportation. First, when CIDCO built the railway link over the Thana Creek, connecting New Bombay with Greater Bombay by rail, a major part of the investment was to be recovered from the operation of the bridge. Consequently, since 1992-93 CIDCO contracted private organisations to collect a toll-tax. Every vehicle, public or private, that crosses the bridge in any direction has to pay a certain fee. Two-wheelers for example pay Re. 1 and cars Rs. 5 per passage, an amount which is not really low but affordable for someone owning such a vehicle. However, public buses have to pay Rs 15 per passage, an amount which is fully recovered from a surcharge on bus tickets, which is thus an extra cost for the public. The same system is in operation for trains, and a surcharge is to be paid on every train ticket for a Bombay-New Bombay travel, ranging from 17% (on a single ticket) to 33% (on a monthly card ticket) on top of the normal price.[40]

Secondly, train stations in New Bombay have been planned and constructed at the edge of the nodes, at a significant distance of the residential sectors. For those people who do not have a bicycle, a motorcycle or a car waiting for them at the station, it is rather difficult and time-consuming to walk from the station to their homes. The only possible alternative is to take a privately run autorickshaw, which charges a minimum of Rs. 5.50 plus metre charge. For those who do have a personal vehicle waiting for them at the station, transportation is still not free, since the parking lots at the train stations, which have been well planned, are charging the customers. A person working in Bombay and leaving his scooter at the station on a daily basis will spend some Rs. 60 per month just for this parking service. In this example, this extra cost increases the cost of a monthly train card ticket with about 40%.

Other elements contributing to the cost of living in New Bombay

Apart from the spurt in land and housing prices, and the high cost resulting from the privatisation of certain urban and social services, living in New Bombay is also relatively expensive in many other ways. Let us first briefly look at the cost of basic urban utilities.

Since the NTDA has, in the initial phase of development, neglected to arrange its own water resources, it now has to buy water for domestic and commercial use at a high cost (Rs 3.50 per m³ in 1996) from the MIDC. As a result, households in New Bombay presently pay the highest rates for water consumption in the region: domestic use of water was in New Bombay in 1994 charged at Rs. 2.80 per cubic metre, whereas in Greater Bombay this was only Rs. 0.60, in Kalyan/Dombivali Rs. 2.00, and in Thana Rs. 1.30 (other figures have been included as an appendix). Since then, water rates in New Bombay have further increased to Rs. 3.65/m.[41]

**Figure 8.11: Water rates (domestic and commercial) in New Bombay
(Oct 1994), compared to three adjacent urban regions**

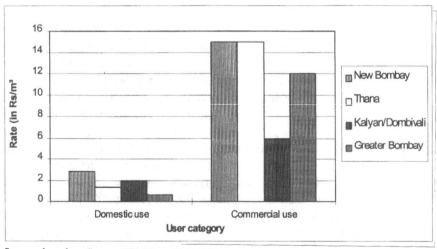

Source: based on figures CIDCO and MWSSB

So far as the cost of electricity is concerned, households in New Bombay are not disadvantaged compared to other municipalities in the region. Apart from South Bombay (provided by BEST) and the Western Suburbs (provided by BSES), electricity rates are the same all over the state of Maharashtra, where the MSEB supplies the electricity.[42] Non the less, the cost of electricity is relatively high everywhere in the state, and the latest hike in tariff rates came in June 1996, when rates for households were increased with 17.5%.[43]

Since in the absence of a municipal corporation for New Bombay, it has always been the NTDA that has provided the usual municipal services to the public, households have always paid so-called 'service charges' to the NTDA instead of the usual municipal taxes.[44] However, since the formal establishment of the NMMC in January 1992, households in New Bombay (in the nodes, in the villages and in the BUDP areas) have been charged a high property tax (of 23% for residential use and 35% for commercial use); this notwithstanding the fact that at the same time they are still paying service charges to CIDCO and that the NMMC has not provided any services yet.[45]

Thus, households in New Bombay have also suffered from relatively high charges for urban utilities and from the existence of a double taxation. Add to that the general inflation in the area, primarily in food products, and it will become clear that living in New Bombay has become a fairly expensive affair for a lot of households.[46]

New Bombay's homogeneous economic base

Apart from CIDCO's general development policies and the working of the land and housing market on the one hand, and the high overall cost of living in New Bombay on the other, the relatively homogeneous economic base of the new city is another factor that partly explains New Bombay's upper middle class profile.

Of the total employment that has been generated in New Bombay since the 1960s (about 150,000 jobs, nearly 70% of which in the secondary sector and 17% in the services sector, see chapter 6), the increase in the supply of unskilled labour has been lagging far behind the increase in demand for unskilled labour. None of the major job providers in New Bombay has been very effective in absorbing a significant number of such workers. The major industries in the TBIA and Taloja are capital-intensive industries which mainly need skilled workers and engineers, and very little additional employment is possible in the saturated TBIA. The same is true for government and semi-government companies such as the ONGC operating in and around the JNPT. The wholesale markets do provide some low-wage employment, but till present, activities and employment in these markets were far below target. Finally, the number of unskilled and semi-skilled workers that can be engaged in service jobs in private or government offices is obviously also very small. Consequently, those unskilled workers in New Bombay that do not have permanent employment are mostly engaged in some petty business, or are continuously trying to get temporary contract works.

If New Bombay was to become a city for all, it is obvious that the employment base was to focus predominantly on the majority of unskilled and semi-skilled workers. They should have been the main target group. Unfortunately this has not been the case, and in terms of economic base New Bombay has increasingly become a city of services and capital-intensive industries.

Conclusion

New Bombay's current population is predominantly an upper middle class population. Even in those areas of New Bombay that were specifically planned to accommodate low-income households (low-income sectors and sites and services), the actual numbers of such households is insignificant compared to what was planned. In such low-income areas, about three quarters of the original households have been replaced by higher income groups. This phenomenon is not recent but has become much more explicit in the course of the last decade.

In an attempt to identify the major factors that are responsible for this evolution, we have mainly been looking at the role and the policies of the NTDA. It is evident that the planning agency is well aware of the existence of such a displacement process and the consequences it has for low income households. Unfortunately, it has done very little to prevent this evolution, and several of its policies, such as for instance its management of the Floor Space Index in low-income areas, have directly contributed to this process. This remarkable attitude poses some serious questions, which will be addressed in more detail in chapter 9.

To come back to the point from where we started, it can be concluded that New Bombay has indeed become a city where 'the man in the street' would *like* to live. The crucial question however is to what extent he still *can* live in this new town. A fairly simple and curious contradiction seems to persist: on the one hand, affordable housing is provided to those low-income households who do not have a monthly income above a certain maximum. On the other hand is it quite impossible to live in New Bombay with a budget as limited as the official levels of eligibility. A much larger household budget is absolutely necessary to purchase a subsidised plot or house, to pay the monthly instalments of the loan, to construct the house, and still be able to cover all the other 'normal' expenses such as taxes, service charges, health facilities, education, and transportation.

Notes

[1] In India the definition 'the common man' is, as in most places, frequently used in an arbitrary way. Sometimes it is defined by referring to specific professional classifications, at other times it is used by the middle class to refer to their own class. In this chapter however the term is mainly based on indicators of income distribution: if in 1989 72% of India's total urban households was earning less than Rs. 2,083 per month (Statistical Outline 1994-95:31), this majority of the urban population should be regarded as 'the common man'. The terms and descriptions used in several planning reports of the early 1970s as well as in the NBDP indicate that the state government as well as the New Bombay planning authorities have been using the term in this latter meaning.

[2] The number of households per node that was included in the questionnaire is 6656 in Vashi, 4219 in Nerul, 2034 in CBD, 2282 in Kalamboli, 2125 in New Panvel, 243 in Koparkhairane, 544 in Sanpada, and 1624 in Airoli.

[3] This resale figure, however, is not worth much, as it does not reflect the resale activity as it is in practice. The same can be said for rental prices. We will return to this at a later stage in this chapter.

[4] The percentage of households that did not mention their income was 2.64%.

[5] Another indicator of household wealth, which at first sight looks fairly contradictory to the general pattern, is the ownership of private vehicles. It is indeed rather surprising to find that 68.7% of all households do not own any vehicle, that just over 11% of households own a scooter and another 11% a bicycle, and that about 6.5% of households own a car or a jeep. It must be noted however that the large majority of households in New Bombay nodes are young families, with 87% of the population below 44 years of age. In the node with the largest group of people above 45 years of age (Vashi, 17% of the population 45+) the ownership of a car or jeep is 10%.

[6] In 1987, the average household income in Airoli node was only 50% of the New Bombay average; today the average in Vashi is significantly higher than the New Bombay average.

---

[7] The rental values given do not correspond with the rates as they are in practice (see further).

[8] Here again the same problem occurred, as 15.9% of females did 'Not Mention' an occupation, probably because many are not working or are unemployed.

[9] In India it is common practice that some public sector employees are being provided with government accommodation.

[10] The 19,207 figure is the total number of households in the sample (19,727) minus the 520 households who did 'Not Mention'.

[11] For an income distribution of households in public and private housing, see figure 8.10.

[12] The figures are based on a sample of 5,470 households, and were originally compiled by the Operation Research Group for the BMRDA. Note: the lowest income categorie that was used was Rs. 700, and not Rs. 750 as in New Bombay.

[13] An estimate of the NTDA's Chief Economist.

[14] DRS-1987 (see further).

[15] Sector 2 of Vashi was selected because it was one of the first sectors in New Bombay exclusively reserved for the Economic Weaker Sections (EWS) and because it is currently located directly opposite New Bombay's major commercial area; sector 10 of Nerul was selected because it was developed about a decade later and because it is one of only two sectors in this node which were to cater predominantly to the EWS; sector 2 of CBD Belapur was selected because it was reserved exclusively for the Low Income Group (LIG), and because it was located at a walking-distance from my own place of residence in New Bombay and therefore conveniently placed for frequent visits and good contacts with the residents.

[16] Included in this figure are graduates and graduate students, nurses, graduates from technical institutes, graduates from law schools, engineers, architects, and post graduates.

[17] Not much can be said about female employment as the data are based on few cases.

[18] Number of houses for each category is estimate based on the size of the houses.

[19] At the time of the survey not many people were living in the BUDP area of Koparkhairane, first because developments here have simply started much later, and secondly because of delays due to a major corruption scandal with the computer draw.

[20] More than 50% of males and more than 90% of females did for example not mention their individual income. This may be because they simply did not want to mention this to strangers, but it is more likely that part of them simply have no official income (as a seperate category 'not applicable' was again not included).

[21] The initial plan was to do this by using questionnaires, but since we already had knowledge of a significant displacement process of low-income households in the nodes, a large part of which was illegal, it was expected that questions on first

ownership, resale methods, income categories, etc. would not get correct answers from the respondents. Alternatively, as the SES-95/96 had not started by then, some of my questions were supposed to be included in that questionnaire. Unfortunately, for reasons which are not entirely clear, in practice very few were indeed included.

[22] Several of these families can now be found in informal housing in the MIDC area opposite Airoli node.

[23] In those days, one truck of sand costed about Rs. 1200, and a 50 kg. bag of cement Rs. 120. Labour costed about Rs. 75 to 100 per day for a skilled labourer (or Rs. 40 to 50 for an unskilled labourer).

[24] The rates for water (supplied for four hours a day) are fixed at Rs. 30 per month, and the connection is for free. The cost of electricity consumption is similar to the ones in the nodes (Rs. 0.90-1.25 per unit), and the connection to the system is heavily charged (from a 'non-refundable deposit' of Rs. 700 earlier, to Rs. 2,700 since a couple of years). The service charges are Rs. 8 to 11 per month, and the property tax that has since recently been claimed by the NMMC is Rs. 168 for a non-built up plot, to be paid in one instalment. The same confusion and anger exists here because of the dual taxation system (service charges by CIDCO and property tax by NMMC).

[25] Finding out where exactly these people moved to is obviously not a simple matter. It was tried to trace them down and to have an interview with them on why and in which circumstances they had left the BUDP areas. Several slums in the Airoli neighbourhood have been visited for that purpose, but only a limited number of the households in these slums had originally come from the BUDP area.

[26] It once happened that such a 'random' computer draw gave an almost 100% Muslim population.

[27] With the hire-purchase formula, lower income groups have to pay the house back within a period of max. 20 years.

[28] Although not officialised, there seems to be an unwritten rule that a large amount of money (up to 50% of the balance between the original price of the house and the resale price) is usually required by the NTDA to actually get the transfer legalised (from an interview with a senior officer working for a housing finance corporation who recently bought property in New Bombay).

[29] An example: a house which five years ago was originally bought for say Rs. 110,000 may now be resold for Rs. 400,000. Either the deal is done unofficially, and then nothing is being asked or declared to CIDCO (which means that according to the official documents the first owner of the house is still the rightful owner), or the deal is done officially, with transfer of the house in someone else's name. In the latter case the documents may only mention Rs. 150,000, so that the transfer charges, as well as service charges and property tax afterwards will be reduced.

[30] Some inaccuracies in the SES-data on housing occupation and resale have come to the surface. One example: in the overall resale figures for all the nodes, the resale figure for Belapur node is 0. In the nodal report for Belapur the resale figure is 22.22%.

[31] Finding a significant number of real estate agents in order to make this survey fairly reliable proved to be no problem, as the job of estate agent in New Bombay is an extremely popular one. Only for Vashi the SES-95 gives a figure of 127 estate agents.

[32] Prices are dependent on location (interior of sector or commercially more interesting major road side).

[33] B-10 type houses are being sold about three times more frequently than F-type houses; A-type houses are being sold about three times more often than B-10 type houses, etc... As we have seen in chapter 6, since the days that the interviews were taken the land and housing markets have experienced a serious downward correction.

[34] Averages from six real estate agents witin these sectors.

[35] Naturally many people talked about bribing the officers to keep their business going without too much obstruction.

[36] In New Bombay's rental market an 11-month deposit has to be paid, which is an additional problem for many families.

[37] (a) CIDCO's estimate is that the cost of most jobs now executed by private contractors is about half to a third of what it would be if it had to do the job with its own permanent staff; (b) such labour trouble has proved to be a serious problem for many urban development agencies elsewhere in India.

[38] Although there are numerous municipal schools outside of the nodes, in the villages, only primary education is provided.

[39] A friend's experience with such a 'charitable' hospital was that when his wife had to give birth to a twin and was in hospital for a month, the household received a bill of around Rs. 10,000 (i.e. close to the Indian annual per capita GNP).

[40] The 75-minute train travel from Belapur node to VT station in Bombay is costing Rs 17 return for a single ticket and Rs. 150 for a one-month card, of which Rs 3 and Rs 50 respectively are surcharges for use of the bridge.The amount of surcharges was proposed by CIDCO and accepted by the Railway Board of the central government.

[41] By mid-1994, New Bombay was having about 60,000 connections to the water network. Of these, less than 6% (about 3,500) were charged according to the metre, and most households were thus paying a fixed price on a monthly basis. In the past, metered charging of water consumption in New Bombay has given numerous problems, for which reason CIDCO increasingly used a fixed water rate. Supply of water is secured only 2-3 hours in the morning and 2-3 hours in the evening. Total supply of water in 1991 was 100 million litres per day, which has increased to 150 million litres per day in 1996.

---

[42] Households pay Rs. 0.60 per unit (of 1 kwh.) if the monthly consumption is below 30 units, Rs. 1.05 per unit for the next 20 units, Rs. 1.05 per unit for all units if consumption is between 50 and 170 units, Rs. 1.60 per unit for all units plus fuel consumption if consumption is between 170 and 400 units, and Rs. 2.30 per unit for all units plus f.c. if total consumption is over 400 units.

[43] *The Asian Age*, 26 June 1996. At present, the electricity bill accounts for almost 30% of production cost in industry.

[44] Since April 1992, the service charges that are imposed by the NTDA are Rs. 7.50 per m² of built-up area for a CIDCO built tenement.

[45] An attempt was made to compare municipal tax packages of six municipalities in Maharashtra with the level of service charges and property tax in New Bombay. These packages however are far from uniform, and a proper comparison with New Bombay proved to be rather difficult.

[46] The rapid inflation in food products is a result of the fact that hardly any food is still locally produced, mainly due to the loss of agricultural land and to soil and water pollution (see previous chapter).

PART III

CONSIDERATIONS ON PLANNING CONCEPTS AND PLAN IMPLEMENTATION

9 Considerations on Planning Concepts and Implementation

Introduction

In this chapter we shall look at four broad items. First, we will discuss the general logic of the original twin-city concept and raise the question whether the development of the area east of the old city on the mainland was the most logical policy choice at the time. Secondly, we will briefly go into the debate on the motives for the creation of a twin-city. Thirdly, we will further critically discuss some aspects of the initial Development Plan and the general planning concept, both in terms of physical planning and socio-economic planning, and their implementation. And finally, we will assess the extent to which New Bombay can be considered a 'self-contained city', as was planned, as well as the overall impact of the new town on the urban situation in Bombay.

Much of the data and arguments in this chapter are based on qualitative interviews with key-people in different sectors of private and government circles. Evidently, at many instances the arguments that were put forward have been contradicting each other, depending on the source and the functions and interpretations of the respondents. Where felt needed, different views on both planning and urban development will be given. Further, the final part of this chapter, discussing the new town's independence and its impact on the city of Bombay, will partly be based on a limited set of data included in the SES-95/96.

The logic of the twin-city concept

When in 1973 the Government of Maharashtra accepted the New Bombay plan, it implicitly recognised that an urban expansion eastwards, onto the mainland, was already in those days more preferable for the future of the city of Bombay than a further expansion northwards, as had been the traditional (unplanned) city development pattern since the 17th century. However, whether this choice for a twin-city was the best possible solution at the time is often questioned.

The arguments put forward by the twin-city advocates, actually not more than just four influential people - Shirish Patel, Charles Correa, J.B. D'Souza and V. Srinivasan - were mainly inspired by overall cost and planning. It was believed that a further urban expansion northwards would cost an enormous amount of money due to the large-scale land acquisitions that would be required and the dearness of the land. Given the undeveloped state of the land to the east of the island on the mainland, acquisition costs here would be much less, and it was believed that even in view of the heavy investments needed for the development of the area, in total the cost of moving eastwards would not be higher than moving development further northwards. Beside this financial argument, the twin-city advocates also believed that too many problems were caused by the specific narrow shape of the island running north-south, especially in terms of traffic congestion.

There can be no doubt that the area on the mainland was bound to be developed at some point. What is sometimes questioned however is whether the development of this area did not come twenty to thirty years too early. Nigel Harris, to name but one influential voice, strongly argued against such a move eastwards at that time, and strongly recommended the development of the large vacant areas of land in the northern and north-western suburbs of Bombay first, mainly for economic reasons (economies of agglomeration). The transportation argument against a further development of the northern Bombay Suburbs was not considered very convincing by Harris. If New Bombay was to become a self-contained city having its own extensive employment base, the same would be possible for any new areas to be developed within the city limits. New residential space in the northern areas would be joined by new local employment bases, which would prevent an even greater pressure on the north-south transportation axes.

Nevertheless, the development of the vacant areas to the east of Bombay was, even in those years, probably the best long-term solution for the city. Throughout Bombay's history there has never been any significant plan implementation. Most plans and development concepts that were produced in the drawing rooms were never really implemented in practice, not in the least because of the traditionally strong links between sections of the state's political class and the city business class, joined in more recent times by the 'Bombay Underworld'. Therefore, even if at the time a further development of the northern areas of the city would have been the best overall solution in theory, in practice things would have never developed in a planned way, and transportation and other urban infrastructure would have come under even more pressure than they are today. Urban development plans carry a great potential, but if the political reality of a city is ignored, they become fairly useless.

There are several more arguments in support of development of the area that is now New Bombay. First, already in the early 1970s the old port of Bombay, located in the most busy area of South Bombay, had to handle an enormous amount of traffic, and the limited space in the area was to a large extent used for port activities and heavy traffic to and from the port. Further expansion of the port was severely restricted. In New Bombay, on the other hand, there was unlimited space to build a new port with all modern facilities. Secondly, from the 1970s, large-scale industrial expansion within the boundaries of the largest cities had been prohibited, and a natural spill-over of industries crossing the northern and eastern boundaries of the GBMC could be expected, and was even desirable. Thirdly, the general concept of the Development Plan for Greater Bombay (1960s) was completely out of touch with reality. Even in those days it was clear that urban development in Bombay could not be limited to just the administrative boundaries of Greater Bombay, and that proper long-term planning had to focus on a much larger area than just the area within the GBMC. An overall large-scale planning and development concept, which was already urgently needed in those days, was chronically lacking. Last but not least, the argument that the New Bombay area could have been developed say twenty years later, does not make sense. Given the extremely rapid growth of Bombay there can be no doubt that in the period in-between most of the land in the area would have been taken over by private developers, and that full acquisition of these lands twenty years later would have been extremely costly, if not impossible. This is exactly the scenario that has unfolded in the Vasai-Virar region, to the north of Greater Bombay.

Debate on the motives for the creation of a twin-city

Over the years, a debate has evolved over the question why the BMRPB, in the early 1970s in charge of urban planning, suddenly and drastically changed its views from the Development Plan for Greater Bombay to the acceptance of the twin-city plan.

One argument that has been put forward (Verma 1985) is that the entire New Bombay idea was created by and for Bombay's business class.

When further expansion of large-scale industries within the limits of the largest cities was no longer allowed, the industrial class of Bombay, according to this view, was given the opportunity to buy large chunks of very cheap and vacant undeveloped land in the newly developing industrial area in present-day New Bombay, at a stone's throw from Bombay. Soon afterwards, the state government started developing that same area with all the necessary industrial infrastructure, after which the New Bombay project was formally accepted. To what extent this argument has real value, or is the product of mere speculation, is debatable. It is a fact that the twin-city idea was originally developed and advocated by a number of people from the private corporate sector, and that it was basically by lobbying and under pressure of these people that the BMRPB ultimately changed its earlier views and accepted this grand new project. What may reinforce this argument is that, looking back, it can be claimed that ultimately it is indeed the business class (industrial class and building lobby) that has gained the most from the development of New Bombay in general.

Different and more nuanced arguments have been put forward by (among others) the major protagonists of the New Bombay plan themselves.[1] They gave the impression that, in those days, the entire New Bombay idea was mainly inspired by a mixture of planning motives, social motives and personal motives, of which the latter were probably at least as important as the former two. At the time, the job of professional urban planner did not exist, and everything concerning urban planning was characterised by a high level of amateurism, even within such 'professional' planning organisations as the BMRPB. According to all four respondents, their major motive for trying so hard to convince policy-makers to accept the New Bombay plan, was because they were absolutely convinced that in terms of physical, social and economic planning it would be the best long-term development strategy for Bombay in general.[2] In addition, however, personal excitement and enthusiasm with the possible opportunity and challenge to create a brand new city from scratch seems to have been equally important. Again and again Mr. Srinivasan tried to convince the politicians to reject the proposed Development Plan for Greater Bombay and to accept the twin-city plan instead, mainly because he was so excited about the plan.[3] Charles Correa from his side also spoke of 'The *brave new world* idea', 'the great enthusiasm in the beginning', and for JB D'Souza 'it may have been more the excitement of a grand new project that made us so favourable for the idea'.

Whatever the real motives were, the fact is that the revised views and proposals of the BMRPB were accepted by the state government. The latter's policies were thus quite contradictory in nature. On the one hand the state government strove for a de-concentration of activities away from Bombay, whereas at the same time it was now planning a large new metropolis exactly next to the old one. The NTDA, obliged to support the state government's policies, was caught in-between the general industrial location policies and the New Bombay proposal, and therefore made it explicitly clear, at several points in the planning document, that New Bombay was not to be developed as an area where new industries could be concentrated. The NTDA's intention was 'to cut back very sharply on the growth of new industrial jobs, diverting such jobs to other parts of the state' (CIDCO 1973:15). When the planners of the NTDA were clarifying that 'only such industries as have a powerful economic reason to be located in this region should be located here' (Ibid:10), they were mainly thinking of the chemical industries, which preferably need a sea-side location.

Consequently, more contradictions come to the surface. First, if the purpose was to limit industries as much as possible, why then was no less than 16% of the land reserved for industry? On the other hand, the NBDP stressed that the new town, in order to decongest Bombay, was to become a self-contained and independent city, and not another satellite town. To develop into such an independent city, however, would require a strong and diversified economic base with a large employment potential. And as it was the intention of the NBDP to divert in-coming migrants away from the island and to become self-containing, it is questionable how this was to be done if no additional employment in industries was to be allowed. What was to become the employment base that would support these migrants and other urban poor that were promised accommodation in New Bombay, if the new town was planned to be based mainly on capital-intensive chemical and port-related industries and on political activities? One of the very few fields in which they were to be absorbed was in domestic services, but such activities would provide them with a wage too small to live in proper accommodation (as the situation in reality has proved). In short, the lack of a clear view on the future employment base for New Bombay was a serious obstruction in planning the destiny of the new town.

Critique on the implementation of the New Bombay Development Plan

As the research results of part II indicate, overall implementation of the New Bombay plan has been far from successful. Whereas the physical planning and implementation of the project could be called relatively successful, in terms of socio-economic and cultural development the project must be considered a failure.

Physical planning and implementation

So far as physical development is concerned, New Bombay's performance is quite impressive. Whereas twenty-five years ago the area was completely undeveloped agricultural land embracing 95 villages, New Bombay has grown substantially and will shortly become a 'metropolitan city' of over one million population. Within a relatively short time-span the NTDA has successfully handled an enormous amount of construction and infrastructure work, and compared to other Indian cities the new city is modern and reasonably clean, is rather spacious and well structured, and is having some good transportation axes.

However, even the physical planning and implementation of the new city is not free of critique. A first major point of critique is the very location of the new city. As can be seen from figure 6.2 for instance, the northern part of the new city runs north to south with the coastal area reserved for residential use. On the eastern side of the residential areas runs a major highway, which separates the residential areas from the huge industrial TBIA (2500 ha.). To the east of the TBIA are the green hills, covered mainly by forest and woodland. The result of this planning concept is that New Bombay's population at present heavily suffers from both traffic pollution and industrial (mainly chemical) pollution. The hills to the east function as a natural curtain, obstructing the gases from drifting eastwards. Moreover, most of the gases from the chemical factories in the Chembur area of central Bombay (popularly known as the 'Chembur gas chamber') also drift eastwards with the predominantly west/south-western winds. These gases too are obstructed by the New Bombay hills and add to the local emission and pollution. It is believed that presently over 20% of young people in Vashi node suffer from severe asthma and other bronchial problems.[4] Official reports of the Maharashtra Pollution Control Board indicate that the pollution levels are still within the prescribed limits, but it is generally believed that large-scale corruption obscures correct measurements.[5]

The chemical factories in the TBIA are not only largely responsible for the air pollution, but also for the pollution of the waters. Many village wells are now contaminated, and the waters in and around New Bombay are so polluted that a lot of the fish that is being caught is either very smelly and not suitable for consumption, or already dead. Moreover, these chemical factories are also a major source of potential disaster, and a serious chemical accident, such as the one that occurred in Bhopal in 1985, is not at all impossible.

At present, industry and the authorities accuse each other for the current situation. Since at the time of planning of the new town the TBIA already existed, this situation is largely the responsibility of the planners of New Bombay, but also of the current policy makers who do not seem able (or do not want) to act against breaches of the environmental regulations. On the other hand, the industries are to be blamed for almost openly setting at naught these rules and regulations.[6]

Well aware of the potentially unhealthy situation and the potential danger for human lives, in earlier days the NTDA took some precautions and developed for example numerous green spaces. However, as time passed, much of the greenery has been sacrificed to commercial pressures, and several places which formerly were parks or gardens are now built-up areas that were sold to the private sector. Furthermore, in the original plan it was intended to line the major roads with trees, to construct separate bicycle lanes, and so on. Nothing of the sort has been put to practice. The major Thana-Belapur road has grown into an extremely unhealthy and unsafe transportation axis, for both the users of that road and for the local residents.

A lot of improvement is also urgently needed in terms of the availability of urban utilities. Electricity and water are being provided to all the citizens of the nodes,[7] but the actual availability is highly restricted: like in Bombay proper, power cuts are a daily returning phenomenon (for which of course the NTDA is not to blame), and tap water is being provided to the public for about two hours in the morning and two hours in the evening on average. A semi-closed sewerage network has been constructed in all of the nodes, but in many places the system is also being used for garbage disposal purposes, and hence has become the perfect habitat for rats carrying all sorts of diseases.

Another major critique are the frequent changes in the land use plan. If the current land use is compared to what was planned twenty-five years ago, the differences are striking. Most often these deflections have been the result of a lack of development control. Many areas which were initially planned as residential areas or were having a social purpose have gradually been transformed into commercial areas. Mostly such transformations first occurred in practice, after which the situation was officialised by the NTDA, and the new areas were formally accepted for commercial use. Take Vashi for example, where the local Sports Complex, which was initially developed on public land, was largely sold off to two private schools, whereas the sports complex itself was privatised and now works with membership fees. At present, only a very small part of the fairly large area is still available for the general public. At walking distance from the sports complex are sectors 7 and 8, which originally were planned as green zones. A major part of these two sectors, conveniently located along the Thana Creek water, has now been transformed into an exclusive residential area, housing some of the most affluent families of New Bombay. In Turbhe node, to take one more example, the township's main park has also been sold to the private sector, and the original park land is now fully built-up. Furthermore, as has been discussed in the previous chapter, residential areas which were initially planned for low-income groups have later been officially transformed (or attempted to be transformed) into high-income residential areas.

Finally, that 'New Bombay has almost no existence of slums' as the NTDA proudly argues (Gill et al:45) is unfortunately a conscious manipulation and/or ignorance of the real situation. The NTDA and other authorities in the New Bombay area (MIDC, JNPT) have established a reputation for keeping the new city slum-free, but only because such unauthorised settlements are regularly removed by force (bulldozing), and because the 60,000-odd people living in such slums in New Bombay are officially not considered as residents of New Bombay. They are completely ignored, are concealed as much as possible, and are not included in any statistics. Officially, they simply do not exist.

Photograph 9.1: Removal by force of illegal (i.c. commercial) settlements (foreground) and squatter housing on MIDC land in Turbhe

Socio-economic planning and implementation

New Bombay has been growing and developing very rapidly over the last ten years, in physical as well as (partly) in economic terms (see chapter 6). However, the results of the field survey indicate that the implementation of the socio-economic chapter outlined in the NBDP has been rather disappointing (see chapters 7 and 8). The population in the villages of New Bombay, the population in the sites and services of the BUDP areas, and the majority of lower income households in the nodes have all been adversely affected by recent developments in New Bombay. Even though the recent developments in New Bombay must be considered a direct result of the impact of the market forces, hereunder we shall argue that ultimately it is the NTDA (and indirectly the state government) that is to be blamed for this evolution. It has clearly failed in what should have been one of its main duties, viz. to keep tight control over developments in the area.

Let us first further evaluate developments in the villages of the new town, and then further discuss in greater detail the developments in the nodes and in the sites and services of New Bombay where, due to a number of factors, a large-scale displacement process of lower income groups has been observed.

Developments in the villages of New Bombay   In chapter 7 we discussed the impact of the development of New Bombay on the original villages and on the village population. The large majority of local people lost their land, their jobs and their incomes. Only the former landholders in the area administered by the NTDA were given a small financial compensation. On the other hand, very little alternative employment was provided, not by the authorities or public companies, nor by the private industries in the area. Consequently, unemployment and the lack of a proper and steady income have been the major problems for the village population. In the meantime, the general cost of living (for food, education, transportation, medical expenses, taxes and services charges) increased many times, and most people spent their compensation money on meeting these expenses.

Several of the rehabilitation schemes that were initially introduced were soon neglected or abandoned, and the major part of the NTDA's expenses on rehabilitation has gone to investments in school building construction instead of directly productive rehabilitation schemes. Moreover, relatively little was done within the villages, and most of them are now still having problems with such basic utilities as piped water or a properly working sewage system. The GES and 12.5% schemes (see chapter 7) are good initiatives in themselves, but came far too late for the majority of village people. They do not have the necessary finances to buy the plot of land they are entitled to, and/or to build a house within the time limits and according to the strict (and expensive) building regulations. Consequently, a large-scale buying-out process has started, with the villagers selling their plots to the private sector.

If the NTDA had really put its mind to it, many of these negative developments could have been avoided. For instance, considering the many difficulties and the persisting strong opposition of the PAPs regarding the acquisition of their lands and the rehabilitation measures, it can be questioned whether no other formula could have been possible than the ordinary acquisition with financial compensation. Charles Correa, one of the key-persons behind the New Bombay initiative, admitted in an interview that a 100% acquisition of land was then not needed, as it is not necessary today. Even though land acquisition seemed to be the only possibility at the time, today he would go for some kind of land assembly, in which the land is given in exchange for equity shares in the Corporation.[8] Correa believes that the price that was paid for the land at the time was indecently low and should have been much higher. Shirish Patel, a second key-person behind the New Bombay initiative, adds another possible mechanism that could have been adopted, viz. to give the PAPs an alternative plot of land in the area which they could have kept hold on for 20 years, and sell it afterwards. Patel nevertheless believes that 'at the time a financial compensation seemed perfectly appropriate, at rates that were also considered very comfortable'.[9] Non the less, as Gill *et al.* rightly argue, if the current transfer of land under the 12.5% scheme had been implemented much earlier, most probably this would have avoided major problems with acquisition and the consequent project delays (Gill *et al* 1995). From the standpoint of the local people such an arrangement, in combination with certain financial compensations, would no doubt have been an acceptable solution.[10]

In terms of rehabilitation, very little has actually been done. In 1973, several CIDCO documents stated that 5% of the entire project cost was to go to rehabilitation. By 1980, the total cost of the project had gone up to Rs. 2 billion, which would have meant that Rs. 100 million should have gone to rehabilitation. Even ten years later, by 1990, the total amount of money that had been spent on rehabilitation was less than Rs. 60 million, the major part of this on bricks for school buildings.

Logically, employment must be considered the single most important element in any rehabilitation and integration. The fact, however, that there hasn't been a single employment generation scheme for the PAPs since the NTDA's bus company, is a strong enough indication of the genuineness of the Corporation's rhetoric on rehabilitation and integration. Would it have been really impossible, for instance, for the NTDA (or the state government) to include certain conditions for those industries that wanted to establish themselves in the area, to require that a minimum of say 5% of their workforce had to be employed from among the village population? If given a proper thought, the rehabilitation and labour generating possibilities for this section of the community are not as limited as the NTDA has always argued. Rather, the real problem is that rehabilitation has never been considered an important topic, as the following two examples indicate.

The first example is the rickshaw driving sector. Whereas rickshaw driving is one of these activities that any PAP could do if properly trained, and although the NTDA's figures say that nearly 260 PAPs were trained as rickshaw drivers, very few PAPs are actually involved in this sector. All rickshaw drivers are outsiders who have managed to get hold of a so-called 'Public Service Vehicle Batch' from the State Transport Department.[11] If the NTDA had exclusively reserved rickshaw driving to the PAPs (which in collaboration with the state government should not have been impossible), and had given them proper training, and provided them with certain financing schemes to rent or buy a rickshaw, today several thousand PAP households would have had a proper income.[12]

The second example are the quarry permissions (cutting stones). According to NTDA figures, 244 of such permissions were granted to the PAPs. However, in practice things are organised quite differently. Instead of employing these 244 people themselves from among the PAPs, CIDCO granted the quarry permission to an advocate of Belapur, who thereby got the right to employ these 244 people himself. However, the people he employed were not PAPs, as was supposed to be, but people who in most cases were already employed, and thus got a second job. In theory, every eleven months the NTDA is supposed to check the conditions and the implementation of the agreement, but does not do that in practice.

These are just a few of many such examples of the difference there is between what the NTDA claims it has done to rehabilitate the village population, and what is coming out of this in reality, i.e. outsiders ultimately taking up most of the scarce alternative jobs. The same can be said about education. If the 'stipends' that are being given to children of PAPs, even though only from 10th standard onwards, are not even sufficient to finance the cost of travel to and from school, these grants must be increased. That the NTDA's Chief Rehabilitation Officer admits that the level of the grants is completely inadequate, but that she 'hasn't got any complaints about this', is an indication of the doubtful genuineness of the NTDA's rehabilitation claims.[13] If the cost of transportation to and from the technical training institutes is one of the major reasons why people drop out of these special programmes, why then is public transportation not being made free of cost for this category of people? If most schools almost openly demand high donations to get a place at the school, even though this is prohibited by law, why then are such illegal practices, which are a major hurdle to affordable education for village children, not being tackled? These are just a few of the major problems that beset the development.

Similarly, if the NTDA genuinely wanted to physically integrate the villages into the new urban environment by providing all necessary facilities, it obviously could have done that. In some nodes the Corporation is presently getting over Rs. 22,000 per m² for residential plots of land, and around Rs. 44,000 per m² for commercial plots of land (compared to the Rs. 3.7 per m² that was - and is - being paid to the PAPs as compensation). How then can a financial burden be the excuse for doing so little in these villages? According to P.C. Patil, the leader of the PAP organisation, by 1996 only four villages could be called 'rehabilitated', and he concludes that 'in most other villages only plots have been provided, no infrastructure nor jobs...'.[14] Hence the feeling in most villages that they have been completely neglected by the NTDA.[15]

When asking older villagers about their prospects for the future they are, almost without exception, very pessimistic. They have little hope, and fear that the situation will become even more difficult for their children, as job prospects are limited. They also fear for the very existence of their villages, as these are visibly going through a process of rapid change and at some places are even physically threatened by the expanding nodes. In short, *disappointment* and *anger* are the key words to describe the local people's feelings about the past; *fear* and *suspicion* are the key words that describe their feelings about the future.

All in all it seems to be coming down to motivation from the part of the NTDA to really wanting to help this section of New Bombay's population. It is quite remarkable that until 1992 'rehabilitation' was falling under the NTDA's 'Lands Section' department. Only afterwards it was transferred to the Social Services Department. Before this move, the provision of stipends was one of the very few rehabilitation measures from which the local population could directly benefit, and which were actually implemented, even though only to a minor extent. The figures on rehabilitation expenditure that have been given earlier are an indication of this evolution.

Developments in the nodes and in the BUDP areas of New Bombay In chapter 8 we discussed recent developments in the nodes and in the sites and services of New Bombay. The general conclusion of chapter 8 was that lower income households in the new town are increasingly being replaced by higher income households. The clearest and most dramatic expression of this process has been observed in the specially reserved low-income residential sectors within the nodes and in the sites and services in the

BUDP areas. In the latter, where over 97% of the houses had been reserved for families belonging to the Economic Weaker Sections, only about 3% of these houses is occupied by such EWS-households, and just 27% by families belonging to the Low Income Group. No less than nearly 70% of the families currently living in these houses belong to the higher income groups (MIG and HIG). A similar evolution can be observed in all the residential sectors of New Bombay that were planned to accommodate low-income families. We have argued that this process was mainly due to (a) the NTDA's land and housing policies and its attitude in relation to the working of the land and housing markets; (b) the NTDA's policies in the provision, management and maintenance of infrastructure and services; (c) the overall cost of living in New Bombay; and (d) the homogeneous economic base. Therefore, as the NTDA has a large influence on all these factors, the planning agency can be held directly responsible for the large-scale displacement process in New Bombay. On a conference meeting in London in August 1997, a senior officer of the Indian Institute of Urban Affairs (IIUA) simply remarked that such a displacement process is a normal thing to happen, and that always and everywhere lower income groups are being pushed out from the moment the land and housing markets pick up. Is this indeed so? Are these developments in New Bombay just normal and unavoidable? Let us first briefly consider the effect of CIDCO's policies in housing and in the provision and management of infrastructure and services.

As has been discussed in chapter 8, from the late 1980s the NTDA revised its land and public housing policies. Land and housing were increasingly brought onto the market in small quantities and were increasingly (and at present exclusively) sold by tender and by auction. Consequently, market rates for land and housing were pushed up to unprecedented heights, and at present even the smallest plots and cheapest houses in the private market have become far too expensive for lower income households. As a result, lower income groups must put their hopes on the public housing system. However, this system also gives numerous problems and the chance for a low income household to get a public sector flat is rather small. Although the public housing distribution system may seem reasonably fair and efficient, in reality it is rather a nightmare. The number of houses reserved for low and lower middle class households (in objective terms the majority of the population) is very small compared to both the total number of houses that is being delivered and the great demand from this class of people. An offer of 1,000 new flats for LIG households, for instance, easily attracts 50,000 applications. Further, the allotment of flats is done by computerised random draws, even though it has been proved in the past that corruption occurs. Also, the conditions to be eligible for a public sector flat are quite stringent and sometimes difficult to prove. An applicant must have stayed in the state of Maharashtra for the last fifteen years, and he must prove that he belongs to the required income category on demonstrating a wage slip of his employer. Many employers, however, do not provide such slips for reasons of taxation. Finally, the number of people applying for public housing with the sole purpose of speculation is fairly high, and they often run off with a big chunk of the houses on offer. Hence the relatively large number of newly built houses in the government's scheme that are kept lying vacant, waiting for someone to come up and pay the price that is required.

If a private flat can not be afforded and the chance to get a public sector flat through the lottery system is rather small, the alternative is the private rental market. Unfortunately, parallel with the land and property rates, rental values of flats in New Bombay have also increased manifold, without there being any government regulation. Renting contracts are for a period of 11 months, and when this period is expired, the landlord may increase the price of his flat to what he feels is the right price. You either agree to pay or you vacate the accommodation. A flat which in 1991 was rented out for Rs. 1,200, was four years later rented out for Rs. 4,000.

Clearly, the development authority has never been able to control these market forces. On the contrary, through its policies and attitudes it has invited and reinforced the effects of these market forces. It is evident that this process could largely have been avoided, if so desired. First, the rules and regulations that were introduced to prevent a rapid selling-off process of public housing have never been enforced, thereby inviting illegal sales of property and leaving low-income (and often illiterate) households to the clutches of development agents, speculators, and other real estate sharks. Secondly, the NTDA could have kept its public flats in possession and could have rented them out to low-income groups at a reasonable price (which is not the same thing as the 'hire-purchase formula'), as is common practice in western countries. Thirdly, the development agency, in numerous ways, has given the impression that it never really wanted New Bombay to become a city for all classes and all people, as was planned. In the sites and services for instance, the NTDA has been actively involved in attempts to remove EWS-households, and to sell off the property to higher income groups (obviously with huge profits). Further, the NTDA has arbitrarily revised the FSI in certain low-income sectors in which there was a high demand, obviously resulting in a sudden sharp increase in property rates in these sectors, and in an acceleration of the buying-out process. In the sites and services of sector 2 in Vashi for instance, the NTDA at one point tried to buy the entire plot of land back, simply because this sector accidentally appeared to be located on some of the most valuable land in the area, opposite the main (but unplanned) commercial centre of Vashi. In short, as can so often be observed elsewhere, in New Bombay also the policies of the government seem to have followed the effects of the private market, *in casu* in land use.

All this raises a number of questions. First, it could be questioned what exactly the purpose of a development plan is, when it is systematically modified following commercial motives. Secondly, I am not at all convinced that a displacement process, as observed in New Bombay, should be considered as an 'automatic' and 'normal' evolution of urban development, not in New Bombay nor anywhere else for that matter. If a public planning agency does not succeed in protecting a large section of the population from the market forces, it clearly fails in what should be one of its most important tasks. From the late 1980s, the development authority in New Bombay seems to have forgotten its specific role as a *public* agency and has been thinking and acting as a private firm and according to private sector principles, putting commercial interests before anything else.

Besides housing, the NTDA's policies in providing, managing and maintaining the urban infrastructure and services have added to the large-scale displacement process of lower income households in New Bombay. As has also been discussed in chapter 8, the development authority's initial policy of planning, developing and maintaining the infrastructure and services itself, was from the late 1980s drastically revised. Basically, the NTDA now only takes charge of the planning and co-ordination of infrastructure works and the provision of services, and the development and maintenance part is increasingly privatised. Crucial social services such as education, public health and public transportation have not escaped from this privatisation drive, with negative price effects for the general public. Here again one could pose the question what exactly the role of a public agency is, when such subjects are left completely to the private sector. Why isn't there a single public high school in New Bombay where children can get good quality education free of cost or at a reasonable price? Why isn't there a single well-equipped public hospital in New Bombay that is affordable to most citizens? Why do buses and trains have to pay to make use of the Thana Creek bridge and is the burden of that extra cost shifted to the public? Why do people who use public transportation have to pay for parking space at train stations? Why is there only private transportation between the train stations and the residential sectors? Taking Belgium, my home country, as an example of the western world, education is here controlled by the government and free of cost, hospitals are heavily subsidised and relatively cheap to the public, nowhere is there a bridge built by the private sector for which people have to pay a toll, a pay system for parking space at train stations is a rare occurrence, and relatively cheap local public transportation is available everywhere. In short, in the provision, management and maintenance of infrastructure and services the public role of the NTDA is also largely ignored. The only thing that currently seems to be important for the agency is making as much profit as possible, even though the NTDA's present receipt balance from the sales of land is impressive, and a lot could be done indeed to relieve New Bombay's citizens from the heavy financial burden they now have to bear.

The overall cost of living in New Bombay is, relatively speaking, very high, and is in part also a result of the NTDA's policies and practices in urban planning. That the price of water for domestic consumption is higher in New Bombay than anywhere else in the region (more than four times the price in Bombay), is a direct result of the fact that in the early 1970s the development authority ignored to arrange for its own water resources. That households in New Bombay are currently double taxed, ones by CIDCO in the form of service charges and ones more by the NMMC in the form of a high property tax, is something quite unbelievable and points to a serious administrative dysfunction. That this situation has existed for over two years now, without proper handling by either the two corporations mutually or the state government as the ultimate authority, says something about the priorities of the policy makers.

Finally, the development authority's economic policies have also played a decisive role in the displacement process, in that the new city's economic base has not been diversified enough to create sufficient employment opportunities for lower income people. It is evident that if New Bombay was planned to be a city for all people, the economic policies should have focused much more on employment generating measures for unskilled and semi-skilled workers. Instead, the economic sectors that have been developed and the kind of jobs that have been created were almost exclusively focused on skilled labour and white collar jobs. Being an uneducated and unskilled person in New Bombay, the best that can be hoped for is to find a temporary job in industry, or a job as a contract worker, because the chance to find a permanent job in the new city is not very great.

Cultural planning and implementation

The main remark in terms of cultural planning and implementation is simply that there has never been a cultural policy, and that not much is being done to vitalise this important aspect of urban life. New Bombay may have become fairly important in terms of population and economic growth, culturally it is completely undeveloped. As Charles Correa, writing on 'The Urban Landscape', says :

> '...perhaps we are paying too much attention to the physical and economic aspects of a city - and not enough to its mythical, its metaphysical, attributes. For a city can be beautiful as physical habitat ... and yet fail to provide that particular, ineffable quality of urbanity which we call *city*' (Correa 1985:81).

Although everyone feels more or less what is meant by 'urban culture', it is a rather vague term that is difficult to describe. Of course, it not only points towards the presence of a thriving cultural life with cinema and theatre halls, music halls, and cafés, but also to a somewhat broader urban concept with the presence of skills, activities and opportunities for all sorts of people, the presence of a large number of city newspapers and magazines, sheer unlimited shopping and entertainment possibilities, etc. In both the narrow and the broader sense New Bombay is undeveloped: whereas Bombay is dotted with cinema halls, in New Bombay there is just one; whereas Bombay has numerous theatre halls, New Bombay has none; well-known rock bands visit Bombay, but will never think of passing New Bombay; the first local newspapers of New Bombay were only recently launched, and so forth. Generally, the perception of people in both Bombay and New Bombay, is that New Bombay is a pretty boring place where never anything exciting happens. Most of New Bombay's people would not even think of their own city to get out for a day of shopping, relaxation and entertainment, but would automatically cross the water to Bombay, where the action is.

Given the fact that New Bombay is almost a 'metropolitan city' in demographic terms, with an ethno-religious mix of people similar to that of

Bombay, the specific cultural flavour and atmosphere that can be felt in Bombay is in New Bombay for some reason absent. Why is this so? Is New Bombay too spacious for the number of people currently living in it? Is it perhaps too clean in a figurative sense (due to for instance the grid pattern of almost identical public housing blocks)? Or is the twin-city simply missing a historical past? An interesting question for debate would be to what extent the fairly homogeneous socio-economic profile of New Bombay's population (although ethnically quite heterogeneous) has an impact on this. In many cities low-income groups such as students, artists or street vendors significantly add to the urban flavour. A city that is largely composed of upper middle class families is bound to be a bit boring, as many upper middle class neighbourhoods in western cities are.

Curiously, very few of the reasonably successful new towns that have been constructed world-wide show a lively urban atmosphere: the atmosphere in the newly created city of Chandigarh is very much different from that in almost every other Indian city; urban living in Brasilia is a far way from the excitement of Rio de Janeiro or São Paulo; and Milton Keynes, the UK's major example of new town development, also shows an unpleasant mix of urban growth and economic development on the one hand, and a lack of excitement and atmosphere on the other. In fact, in New Bombay the two older towns of Panvel and Uran, although together they only have some 60,000 persons, have a much more lively atmosphere than Vashi node, which has a population nearly twice as large.

This lack of an urban atmosphere and culture is probably one of the most important factors in that the state's politicians have never ever seriously thought of shifting a large part of the state government and administration out of South Bombay, and into the newly developed CBD at Belapur in New Bombay. Although such a move was a key element of the Development Plan for New Bombay, partly to relieve South Bombay from its congestion and partly as a major catalyst for development in New Bombay, no doubt it would have unleashed a revolution amongst the bureaucratic classes. In Brasilia it has also taken many years before such a move was forcefully put through, against the massive opposition of the domestic administrators and foreign missions.[16] In the mid-1980s, when New Bombay was not yet really developing, Charles Correa would still believe that the government didn't decide to shift office to the new city because there was a general lack of most things that are needed by an army of bureaucrats and their families: schools, medical facilities, shops, comfortable transportation, telecommunication facilities, entertainment, and so on. By now we know better. Since then, all these things have been provided, but still no one is even considering shifting government offices to New Bombay. Why this is so is probably, as Correa points out, brutally simple: 'the people who can pull the levers to change Bombay *don't need* to change Bombay. In fact they often stand to gain by the status quo' (Correa, 1985:107). Indeed, Bombay's elite, which is partly made up of the city's top bureaucrats and politicians, is having quite a comfortable life living in South Bombay (very much unlike the ordinary citizen), where schools, clubs, colleges, sports stadiums, cinemas, housing, offices and the water front are all located within walking distance. Thus, those who ultimately have to decide on a shift of the state's administration to New Bombay would only make it more unpleasant for themselves. Hence the general reluctance to any change in the current situation, to the disadvantage of the large majority of Bombay's population.

New Bombay: a *self-contained city* with a regional impact?

The ultimate objective of the New Bombay project, put forward in the early 1970s by the Government of Maharashtra, was clearly to reduce population growth in Bombay and to keep the population of the old city within manageable limits. The new twin-city was to divert the continuous stream of in-coming migrants, who would otherwise go to Bombay, as well as to attract some of Bombay's existing population. Consequently, some time later, the newly established NTDA stipulated that one of its principal activities would be to promote a whole range of economic activities (business, office activity, wholesale markets, warehousing, transport, etc.) 'in order to evolve expeditiously a sound economic base for a self-sustained growth' (CIDCO 1986, 1989, 1992). Thus, in order to relieve the old city of Bombay of its most severe pressure, New Bombay was planned to become a new *independent* metropolitan city, rather than merely another satellite town of Bombay which would heavily depend on the old city.

To what extent has New Bombay developed into such a 'self-contained' and independent city, and to how far is it having a significant impact on the urban situation in Greater Bombay? The first question has been approached on the basis of a (limited) set of data included in the SES-95/96. The latter question is a more difficult one, since many different variables (geographical, political, and economic) outside of New Bombay may also be having an influence on the urban development process in

Bombay. Consequently, the results of the latter exercise are tentative, and to a large extent based on personal interpretation.

New Bombay: a self-contained city?

The economic base and the commuting pattern of the local population are probably the best indicators of New Bombay's independent character. In the previous chapter we already argued that New Bombay's economic base is largely focused on skilled and semi-skilled labour, and is therefore rather homogeneous. Apart from being homogeneous, however, the new city's economic base has also been quite small. Whereas in 1995 New Bombay accommodated about 700,000 people, only about 150,000 jobs had been created, out of which nearly 104,000 were provided in the industrial estates only. The total number of jobs in government and private offices and shops was less than 30,000. Consequently, a large number of people living in New Bombay are daily crossing the Thana Creek to reach the major job centres of Bombay.

Data of the first SES (1987) indicate that in those days less than half (47.7%) of the working population living in New Bombay was working in the region, and that 40.1% of New Bombay's workers were daily commuting to Greater Bombay (the majority - 25.3% - to the old CBD in South Bombay). By 1995, the percentage of New Bombay's working population employed in New Bombay had increased to 60.76%, and the percentage of commuters had decreased to 39.24% of the total work force. More than three in four of these commuters were employed in Bombay (SES-95/96:R23). Thus, given the increase in population in New Bombay since 1987, the number of jobs in New Bombay has seriously increased. On the other hand however, although percentage-wise the number of workers commuting to Greater Bombay has decreased, in real terms their number has more or less doubled, from 15,520 to 31,336.[17]

Table 9.1: Commuting pattern of New Bombay workers, 1987 and 1995

	1987	1995
Working in New Bombay	47.7%	60.8%
Commuting to Greater Bombay	40.1%	30.0%
Working elsewhere (Thana, Kalyan, ...)	12.2%	9.2%
	100.0%	100.0%

Based on data SES-95/96

This gradually changing commuting pattern is an indication of New Bombay's growing economic base, and a reflection of the new city's increasingly independent character. Moreover, a number of very recent developments, such as the actual shift of large parts of the wholesale markets to New Bombay, the expansion of economic activity in and around the JNPT port, and the resulting increase in private business and other services, add to the expectation that this gradual development into a full-grown self-contained city will continue in the years to come.

Likewise, in social terms New Bombay's population is also becoming less dependent on the old city. In 1987, not less than 61% of all non-work related trips were going to Bombay and just 25% to one of the New Bombay nodes. The large majority of these trips to Bombay originated from Vashi, Nerul and CBD nodes, and were done for social reasons such as friends or family visits (34%), shopping (21%), recreation (4%), and medical reasons (4%) (SES-87:38). By 1995, however, the percentage of non-work related trips to Bombay had drastically decreased to 9.8%, and nearly 33% of such trips are now going to Vashi, which has developed into a modest shopping town. The main purposes of such trips are still shopping (44.1%), or a combination of shopping, social visits, medical and recreation (30.2%).

That New Bombay is at present gradually developing into the self-contained city it was planned to be, does not conceal the fact that it has taken more than 25 years to reach this point of development. Until recently, the new city could not be considered much more than a semi-developed satellite town, heavily dependent on Greater Bombay for economic as well as social facilities and opportunities. It could be argued in this sense that if the state government would have concretised its initial proposals of the early 1970s to shift major parts of the political and administrative level out of South Bombay and into New Bombay, growth and development in New Bombay would have picked up much earlier, and the new town might have been a fairly independent city for quite some time. That the state government has never made any concrete preparations to shift any major political functions out of South Bombay is regrettable, as it not only would have given New Bombay the much needed growth impulse it urgently needed for so long, but as it would also have significantly relieved South Bombay of much of the economic and urban pressures.

Impact of New Bombay on the urban situation in Greater Bombay

As argued above, whereas the percentage of people commuting from New Bombay to Greater Bombay has come down from over 40% to 30% between 1987 and 1995, in real numbers about twice as many people were in 1995 commuting to Bombay. Consequently, recent economic activity and growth in New Bombay are not having a significant positive impact on the situation in Greater Bombay in terms of pressure on the north-south transportation axes (road and rail) towards the two major job centres in South Bombay.[18]

On the other hand, however, due to the recent shifts of large parts of the wholesale fruits and vegetable markets from South Bombay to Turbhe, and of the wholesale iron and steel markets from South Bombay to Kalamboli in New Bombay, South Bombay has been significantly relieved of heavy load traffic. Moreover, the recent partial shift of activities from the old BPT port in Bombay to the rapidly expanding new JNPT port in New Bombay, also adds to a relocation of heavy load traffic from South Bombay to New Bombay.

As has been discussed in chapter 4, in terms of population growth the Census of India 1991 indicated a clear slow-down in population growth in Greater Bombay, and even a slight negative growth in its southern part. Whereas the overall growth in Greater Bombay was 43.8% from 1961 to 1971, and 38.0% from 1971 to 1981, in the latest decade from 1981 to 1991, the growth of the population in the old city has slowed down to 20.4%. In real numbers, the total population of South Bombay decreased with 110,151 persons to 3,174,889, but in the Suburbs the population further increased, from 4.96 million in 1981 to 6.75 million in 1991, resulting in a total population for Greater Bombay of 9,926,000 in 1991.[19]

However, as mentioned earlier, we cannot simply argue that this relative slow-down in population growth in Greater Bombay is a consequence of the growth of New Bombay. Other variables, such as the enormous increase in property rates and the difficulties in finding affordable accommodation in Greater Bombay, may be of much more importance in this process. It is however safe to argue that part of this slow-down is probably a result of development of New Bombay. Indeed, the data of the SES-95/96 indicate that about 37% of the people currently living in the nodes of New Bombay (about 121,000 persons) are middle-class people who have moved straight into their current accommodation coming from a location in Greater Bombay. They have shifted to New Bombay mainly because of problems with their erstwhile Bombay accommodation.[20] Beside this group of New Bombay residents, another 35.0% of all households have moved to their current house coming from another flat in one of the nodes of New Bombay (for reasons of larger accommodation, accommodation on ownership basis, etc.), and it may be assumed that the large majority of these have also originally come from Greater Bombay. Only 15.6% of the 79,000-odd families currently living in the nodes of New Bombay have, over a timespan of 25 years, directly migrated into New Bombay without first passing Bombay. Of these, 5.3% have directly migrated from some area within the BMR (such as Thana or Kalyan), 4.7% from within the state of Maharashtra but outside the BMR, and 5.6% from outside the state of Maharashtra.[21]

As the number of families that have directly migrated to New Bombay without first passing Bombay has been relatively small, one of the original aims of the New Bombay project, to divert new in-coming migrants away from Bombay, has largely been a failure.[22] Moreover, if the number of people that have shifted from Greater Bombay to New Bombay in the 25-year timespan between 1972 and 1997 is being compared to, for example, the 168,000-odd people that were added to the city of Greater Bombay annually between 1981 and 1991, it will become clear that the aim to attract some of the existing population away from Bombay has only been moderately successful.[23] Without development of New Bombay, the population of Greater Bombay would at present undoubtedly be even higher, but the overall impact of growth and development of New Bombay on population growth in the old city has nevertheless been limited.

Conclusion

Even as early as the late 1960s, planning and development of the area east of Bombay was to be regarded as the best possible long-term solution for the city as a whole. The argument put forward by some, that the creation of New Bombay was really not more than a grand design of a few business people to please the business class, may have some value, but motives of planning and social welfare, as well as personal motives, seem to have been at least as important. The boyish excitement of creating a brand new city seems to have been the prime motive behind the tireless efforts of the New

Bombay protagonists to make the BMRPB change its views and accept the twin-city concept.

Although the physical planning and construction of New Bombay could be called relatively successful, it is not free from critique. The very location and planning lay-out of the new city was rather unfortunate, as severe problems of pollution could have been expected. Further, the new city still suffers from a grossly inadequate provision of basic urban utilities. Insufficient housing for lower income groups in practice, expressed in the relatively large pool of unorganised slum dwellings in the MIDC areas, is another of New Bombay's main physical characteristics.

Culturally, New Bombay is almost completely undeveloped and not much is being done to vitalise this important element of urban life. In a strict sense, New Bombay certainly does not figure on the cultural map of India. In a broader sense the twin-city is, like most newly created towns and cities in the world and in India, rather sterile. *Navi Mumbai* lacks the activity, the imagination and the creativity to get over its reputation of a 'town-without-a-soul', and to become a true 'city'.

Socio-economic planning and plan implementation have been rather unsuccessful. For the people in the villages very little alternative employment has been provided to compensate for the loss of land, jobs and incomes, whereas in the meantime the cost of living has increased manifold. Proper rehabilitation of the people and integration of the villages has never been a prime concern of the development authority and leaves much to be desired. Due to poverty on the one hand, and the working of the land and housing market within a laissez-faire framework on the other, a large-scale buying-out process of land, and even some original village houses, has occurred. Much more should and could have been done in these villages, and if the development authority had really wanted, the faith of the village population at present would not be so pitiful.

The large-scale displacement process of lower income groups in the nodes and in the BUDP project areas is to a very large extent the responsibility of the NTDA, that never seemed to be able (or to be willing) to keep control over developments. Since the late 1980s, the Corporation's former policies in housing and in the provision of infrastructure and services have drastically been revised. At every level this has caused a dramatic increase in prices of land and housing, resulting in massive speculation and real estate activity, and has pushed the cost of several social services up to levels unaffordable to many. At present, the NTDA can no longer be considered as a *public* agency, planning and working in the interest of the people, but rather as an agency slavishly following private market principles and working mainly for itself and its shareholders. The development authority had indeed to be financially self-sufficient, due to which the policy choices have always been rather limited, but the huge profits the agency is currently making could easily allow for a truly public role as well. Therefore, instead of playing the role of giant real estate developer and speculator, after the example of the DDA in Delhi, CIDCO should urgently rethink its role and function as a public sector agency.

So far as the new town's level of 'self-sufficiency' is concerned, things have significantly improved in later years. Until recently, New Bombay was nothing more than another semi-developed satellite town, greatly dependent on Greater Bombay for economic and social services and opportunities. Whereas a major shift of political activities from South Bombay to New Bombay, as was initially planned, would have been a very strong catalyst for development in New Bombay, it has taken more than twenty years for New Bombay's economic base to become strong and diversified enough as to provide a minimum of jobs to the local population. Consequently, the new town can only since recently been considered a 'self-contained city in the making'.

The overall impact of New Bombay on the urban situation in Greater Bombay is, nevertheless, fairly limited. Although the old city has been relieved of a lot of heavy-load traffic, the pressure on the old transportation corridors into South Bombay has, due to an increase in railroad traffic and private vehicles on the road, not been reduced; rather on the contrary. Furthermore, although population growth in Greater Bombay has somewhat slowed down since 1981, the impact of the development of New Bombay on this process is overall not very significant.

Notes

[1] From interviews with Charles Correa (16 Feb. 96), one of the most renowned Indian architects who would later become the first Chief Architect of CIDCO; with Shirish Patel (23 Feb. 96), a civil engineer who would later become the general co-ordinator of CIDCO's planning team; with JB D'Souza (18 Jan. 96), a top- level bureaucrat who would later become the first Managing-Director of CIDCO; and with V. Srinivasan (over the phone, 5 March 96), who in those days was heading

SICOM and was working very closely with the then Chief Minister of Maharashtra.

[2] '... because we saw this map of Bombay in which the BMC had prepared a Development Plan for the municipal area, which showed very carefully worked-out pockets of green spaces, industries, commercial and residential areas and so on.... but only as far as the limits of Greater Bombay. So we asked them what would happen beyond that, and they replied that *that is not our business*. So we said *do you really believe that nothing is going to happen on the other side of that administrative line...* to which they answered *well, we can't do anything about that...* So their planning was confined to the geographical limits over which they had jurisdiction, and they were really not looking at the context of neighbouring areas at all. So it all looked fairly normal that Bombay would not only grow northwards, but also eastwards, where this new bridge had just been opened in 1970' (Shirish Patel).

[3] 'I was fascinated by the idea of a new water front city across the harbour, with café's and all that... It was only fascination that made me so determined to push through this idea in government circles' (V. Srinivasan).

[4] Pollution in Bombay proper is obviously even worse. Lead for example, one of the most poisonous particulate matters, was found to be from 2.5 to 18 times the permissible limits (World Bank). Honorary director and head of the department of respiratory diseases at Grant Medical College and J.J. Government Hospital speaks of 'dangerous levels' and calls the situation 'explosive'. According to him 'the number of asthmatics in the city has steadily gone up and today 50 per cent of the city's population suffers from asthmatic bronchitis' (*The Asian Age*, 24 Nov. 1997).

[5] Since recently, even the local and state authorities admit (off the record) that there is a serious problem.

[6] It is generally known and can be easily observed that the most polluting gases are being emitted at night. Everyone knows it but not much is being done about it.

[7] The situation in the villages is an entirely different matter.

[8] It should be noted that although such a formula would have given the former landowners a form of life-long dividend, it would not have been of much help in the first 10-15 years, as development of the area was not really moving forward.

[9] From an interview with Shirish Patel (23 February 1996).

[10] More recently CIDCO has adopted alternative systems of land acquisition in two other areas for which it was designated the development authority: Waluj and Vasai-Virar. In the latter, so-called Transferable Development Rights (TDRs) have been introduced.

[11] It seems to be general practice that in order to get such a batch one has to bribe ST officers.

[12] Is there a proper economic argument why CIDCO could not have monopolised the entire rickshaw driving business, making even a profit out of it?

[13] Out of the total sample of nearly 49,000 villagers included in the *Village SES-96*, only 464 students received a CIDCO stipend (109 for technical training and 355 for college education). The amount of money that was spent on these stipends was Rs. 633,515, or Rs. 1,365 per student on average (village SES-96).

[14] From an interview with P.C. Patil (29 February 1996).

[15] Due to such feelings of neglect, several villages have organised themselves in the so-called *Sangarsha Samiti*, to follow up their demands at CIDCO Bhavan.

[16] At present, administrators in Brazil shift office every week. In the first part of the week they reside in the new capital Brasilia, after which they move back to Rio (predominantly to be out of Brasilia for the weekend).

[17] The latter is an estimated figure derived from the total number of people in New Bombay commuting to Bombay in 1995, multiplied by four (sample = 25%).

[18] Private traffic, for example, has significantly increased. Whereas in the early 1970s, Greater Bombay was counting 96,000 private vehicles (cars, motorcycles,...), by 1995, their number had increased at a rate of 8 to 10% per annum, and the number of private cars alone was put at nearly 300,000 (34,000 of which are taxis) (Statistical Outline 1995:209).

[19] In South Bombay, the population increased in wards A, F and G, and decreased in wards B, C, D, and E. In the Suburbs, the population increased in all the wards but one (N, which was split), and the increase was most significant in wards R (+416,000), K (+317,000) and S (+290,000) (the latter which now includes part of the population of ward N).

[20] Of these, 12.5% had come from South Bombay, 4.9% from the Western Suburbs, and 19.6% from the Eastern Suburbs.

[21] The remaining part of the population currently living in the nodes were originally living in the villages, in the two old municipal towns of Panvel and Uran, or have moved in from abroad.

[22] For a recent study on migration into New Bombay (the only one so far conducted), see Prasad and Gupta (1995).

[23] It should be noted however that in contrast to earlier years the fairly low rate of occupation in New Bombay is at present not a consequence of a lack of demand from potential buyers, but rather of an insufficient supply of housing.

10 General Conclusions

Introduction

This study started from a number of questions concerning urban development processes in New Bombay, India's largest and most significant urban planning experience. The policies that were introduced in New Bombay have later also been adopted elsewhere in the region, and the impact of the 'New Bombay model' is therefore of regional and even national importance. This New Bombay model was fairly peculiar and unique in that it was based on an ambitious and unusual combination of planning goals and development strategies: on the one hand the new town, which was to be build-up from scratch, was to be geographically located at a very short distance of one of the world's largest cities, whereas on the other it was to become an *independent* 'counter-magnet' of metropolitan size. Furthermore, the land was to be used as the major resource for development and was to fully finance the entire cost of the project, whereas at the same time the new town was also to become 'a city for all', where affordable housing would be available to every household in need of shelter, irrespective of social class and financial strength, and where basic urban services would be provided both cheaply and sufficiently.

The overall objective in this book was to examine to what extent the implementation of the New Bombay project, given these specific goals and development and financing policies, has been successful in terms of physical and economic growth on the one hand, and social development on the other, the latter mainly in terms of the impact of urban planning policy for the urban poor. The major aim of this study was, therefore, to draw some lessons, be they positive or negative, from the urban planning experience in New Bombay and the potential implications for urban planning theory. Furthermore, the study, which was largely based on empirical research, aimed at being a pioneering study on urban development in New Bombay in general, a topic which so far has rarely been the subject of independent research.

Having done this research, it can be concluded that the urban planning experience in New Bombay has, indeed, provided some very valuable positive as well as negative lessons in respect to Third World urban planning and development. By far the most positive lesson is the

271

visible rapid growth of the new town, physically, economically, and population-wise. The New Bombay concept has demonstrated the validity and usefulness of the core elements of the New Bombay model. It has indeed proved possible, even in a developing country with scarce financial resources, to build up a new town from scratch, using the land as the major resource for development, and at the same time being self-financing. Furthermore, the New Bombay example has proved that such new town development can be a valuable tool in the restructuring of the regional urban growth pattern.

However, given the specific context and the specific planning objectives and implementation policies, it would be misleading to evaluate the New Bombay model as a possible general concept for Third World urban planning. In New Bombay some external factors, such as the rapidly increasing property rates in the old city, pushing out middle class households to the periphery, have significantly contributed to the growth of New Bombay. The fact that New Bombay is so closely located to Bombay has been a crucial element in this respect. If New Bombay would have been located at a significant distance, as has been the case with most newly developed towns in the developing world, New Bombay would never have developed in the same way, and would perhaps never have developed at all. On the other hand, the fact that there has been only one physical connection between Bombay and New Bombay has also been of significant importance. Because of this restricted physical linkage between the old and the new city, New Bombay was bound to be self-containing to some extent. In most other places, the development of a new town so close to an existing major city would not have made much difference to the urban growth pattern, but would have been a simple and ineffective extension of the city limits. In short, although New Bombay has been fairly successful in terms of physical and economic planning and demographic growth, the context in Bombay and New Bombay, geographically as well as economically, has been very peculiar. To adopt the same planning policies in another context, or to copy the New Bombay model elsewhere, will therefore not be sufficient as a tool for successful restructuring of the urban growth pattern.

At least as important as the positive contributions of the New Bombay model to urban planning, are the negative lessons to be learned from the New Bombay experience. Socially, if compared with the objectives in the development plan, New Bombay must be considered a failure. The new town does not cater to all income groups, as was planned, and the specific policies and attitude of the planning authority have caused a dramatic displacement of low-income households in all the areas of New Bombay. Furthermore, the performance concerning the rehabilitation and integration of the original village population has also been rather poor, and to a significant extent characterised by neglect.

The main question now is whether this socio-economic shift 'upmarket', which has obviously been a very important element of the urban development process in New Bombay in general, could have been avoided? New Bombay was indeed to become an independent new metropolitan city on a self-financing basis, and the policies that could have been adopted to fulfil that purpose were therefore limited, and were automatically to be inspired by commercial and profit-making motives. But to how far was it unavoidable that the new town would develop into a socially 'upper class area'? There is no reason to believe that this process could not have been avoided. Basically, the 'shift upmarket' was a direct result of active policy and a laisser-faire attitude. In general, therefore, New Bombay could have been much more than just a reasonably well planned and fastly growing city. Given slightly different policies and a different attitude, it could just as well have become 'a city for all', as was planned.

In respect to these observations, a number of important questions come to the surface. To what extent, for example, have the policies of the development authority been autonomous policies, and to what extent had the State government a direct impact on the policy choices adopted in New Bombay? Is CIDCO implicitly or explicitly acting on behalf of certain groups? Has the State, or some of its agents, a direct or indirect interest in the way New Bombay develops? We shall come back to these questions further in this chapter. Let us first briefly summarise and highlight the major points of the study again.

Summary of the major points of the study

Part I of this book was devoted to the broad context of the research topic, viz. the processes and patterns of urban development and the many related problems of rapid urban growth, both on a macro level (the 'Third World' and India) and on a micro level (Bombay). The book started with an overview of urban development processes and urbanisation patterns in the developing world in general (chapter 2). It was demonstrated that the extremely rapid urban growth processes world-wide since the 1950s were almost exclusively the result of urban growth of the developing world alone, with the metropoles of the Third World overtaking those of the developed western world, both in number and in terms of absolute population growth.

As every country, even within the same region, has its own specific history, and its own specific economic, political and cultural characteristics, elements which are to a large extent the direct fundamentals of the urbanisation and urban growth processes, the general approach was narrowed down to a discussion, in chapter 3, of the urban development experience in India. This country has not only the longest experience with urbanisation, but has today also the largest concentration of 'mega-cities', and the second largest urban population in the world. The ambiguous relationship between a fairly modest urbanisation rate on the one hand, and a very rapid urban growth pattern on the other, came out as one of the most striking features of urban development in modern India. Most significant is that in India the urbanisation process, percentage-wise, is in fact only in its infancy, and that despite government attempts to slow-down economic concentration and urban growth, the cities of India will continue to grow dramatically in the years to come. This argument was based on the fact that urbanisation and modernisation are two processes which nearly always go hand in hand. Therefore, the challenge for India will be to find methods that can cope with the expected growth of cities rather than for methods to stifle that growth.

In chapter 4 the focus was further narrowed down to a discussion of the historical processes of urban development in Bombay since the city's inception in the mid-17th century. The growing political and economic functions of the city in the colonial period, largely an outcome of its role as a major support base for the Empire, were the main cause of the demographic growth pattern and the changes in the city's spatial pattern. However, growth in the colonial period was not exceptionally fast. Only after Independence, when the urban economy would become much more diversified, Bombay experienced its fastest growth in history, with the population increasing by about ten times. As a result of this dramatic growth, the city has increasingly been confronted with serious urban problems and inadequacies, and with a very significant level of congestion. In the final part of chapter 4 we have discussed the most significant of these problems, and have given an overview of the debate and different policy proposals that were put forward for the future development of the city and the region at large.

In *Part II* of the book the focus was shifted from urban development processes and urban growth patterns on a macro and micro level towards the urban development processes in the newly urbanising area of *Navi Mumbai*, with the main objective to test the overall success of the New Bombay project after 25 years of project implementation, and to learn some lessons from the new town's development experience. In chapter 5 we first examined the origins of the twin city concept and the birth of New Bombay, as well as the physical and socio-economic characteristics of the area at the time that development was started. Subsequently, the major points and objectives in the New Bombay Development Plan were discussed.

Chapters 6, 7 and 8 presented the major body of the book, based primarily on empirical research. Field research and data collection were designed with the intention to provide as much information as possible on the physical, economic and social developments in the new town, as compared with the major objectives of the project initially put forward by the State government and the development authority in the early 1970s. In chapter 6, a number of physical and economic key-indicators such as population growth, economic growth and activity, land prices and property rates, government activity, and urban and social infrastructure were put forward in order to test to what extent the new town has been growing and developing over the last 25 years. Numerous figures and statistics have demonstrated that since the mid-1980s New Bombay has been growing very rapidly, both in economic and physical terms, and that the population of the new town has increased very significantly in the last ten years, to currently almost 800,000 people. Prior to 1985, however, urban development was sluggish and growth did not really pick up. The physical linking by rail of New Bombay with the major job centres of Bombay on the one hand, and the liberalisation of India's economy since 1991, creating a spurt in the demand for commercial and residential space on the other, were identified as the two major factors that can explain the sudden rapid growth since the mid-1980s. In general, the fact that the development authority has been able to develop New Bombay into a functional and rapidly growing city, having a certain impact on the regional urban growth pattern, is no small achievement, and no doubt the most positive lesson to be learned from the New Bombay experience.

The major concern in chapter 7 was to examine the social and economic rehabilitation of the original village population, and the physical integration of the villages into the developing area. The village survey clearly demonstrated that in order to speak of 'socio-economic rehabilitation' and 'physical integration', a lot more should and could have been done. Numerous rehabilitation schemes (alternative employment, education, technical training, provision of basic urban utilities in the villages,...) were planned on paper, but few have actually been put to practice. The combination of large-scale land acquisition and loss of employment and income on the one hand, and a highly selective and grossly inadequate process of social and economic rehabilitation of the population on the other, have resulted in a general atmosphere of disappointment and anger towards the development agency, and strong feelings of fear and suspicion for the future. The land compensation schemes, implementation of which were only recently started, could have been an important balancing factor for a large group of land-owning households, if only they would have been introduced fifteen years earlier. Generally, one of the major consequences of the poorly implemented rehabilitation and integration schemes is that many village households at present can no longer afford to live in the area, and that urban land developers, property agents and high-income families are gradually taking over the original village land, with the physical outlook of some villages already changing and increasingly showing elements of a semi-urban environment.

Another major objective of the New Bombay plan was to develop the new town into 'a city for all', irrespective of social class, where everyone would be able to find affordable accommodation, and where basic urban services were to be provided cheaply and sufficiently. In order to test the extent of success of this ambitious objective we have, in chapter 8, analysed the socio-economic profile of the population currently living in the nodes of the new town, and compared this with the profile that would be expected on the basis of the planning and housing objectives. Generally, the data clearly indicated that New Bombay's present population is anything but 'the average Indian population'. About four in five households currently living in the nodes are 'higher income families' (MIG and HIG), and just one in five are 'lower income families' (LIG and EWS). This unbalanced household distribution pattern in New Bombay had already been observed in 1987, when the population of the new town, in terms of household income, could be compared with the population of affluent South Bombay. Since 1987, however, the situation in New Bombay has only further polarised, and the proportion of high-income families has further increased.

In addition to the nodes in general, we have also analysed the socio-economic composition of the population in one specific, randomly selected node, and have done the same in certain low-income sectors within the nodes where low-income housing was provided almost exclusively for the LIG and EWS income categories. Significantly, in all of the three examined low-income sectors it was found that the number of middle and high-income families occupying such low-income housing is exceptionally large and does not correspond at all with what was planned. Furthermore, in the sites and services of New Bombay, which were exclusively reserved for New Bombay's poorest households, the same evolution could be observed, with the number of households legitimately occupying these dwellings being relatively small, and nowhere exceeding 30%. In short, in all the three different socio-economic areas of the new town (the nodes, the low-income residential sectors, and the sites and services) a significant displacement process of lower income households has been in operation for quite some time and is continuing.

Naturally, we have tried to identify the major factors behind this displacement process and socio-economic shift 'up-market'. As mentioned earlier in this chapter, the policies and attitude of the development authority have been considered as most important. Its policies in land and housing and its laissez-faire attitude towards the working of the land and housing markets were identified as crucial elements in this process. Further, some of the NTDA's privatisation policies in infrastructure and services, and its increasing attitude towards commercialisation and profit-making, using the urban land not only as a resource for development but increasingly also as an instrument for speculation, have significantly added to the general shift up-market. The relatively high inflation in the area, a fairly natural process in a developing area, has further added to the unaffordability for low-income households to live in the new town, the more since the absorptive capacity for unskilled labour has been far insufficient and has prevented the majority of low-income people from earning a wage high enough to sustain a living in the area.

In *Part III* of the book we have further discussed some particular, interesting elements of the original planning concept. We have argued, for example, that although it may seem that the creation of a twin city was not the most logical policy choice in the late 1960s, the development of the area east of Bombay on the mainland was probably the best long-term solution for the city of Bombay. In this chapter we also further examined the implementation of the major objectives of the New Bombay

Development Plan. Although in terms of overall physical and economic growth, urban planning in New Bombay may be termed successful, it was also argued that the locational choice for the town has been rather unfortunate in environmental terms, and that pollution has become an important factor in urban living. Furthermore, it was found that the new town also suffers from a grossly inadequate provision of basic urban utilities, and that there is a huge gap in this respect between what was planned and what has been implemented. On a cultural level it was argued that New Bombay is still fairly undeveloped, and that very little has been attempted to vitalise this important aspect of urban life. Like many other new towns in India and outside, New Bombay lacks the creativity and urban atmosphere that is often so typical for older cities, and the twin city is rather sterile. Further, in testing the extent to which New Bombay can be considered a self-containing city, having a significant impact on the urban situation in the region, we have argued that up until recently New Bombay was not really more than a semi-developed satellite town, greatly dependent on Bombay for economic as well as social services and opportunities. It took some twenty years before the new town's economic base was strong and diversified enough to provide a minimum of jobs to the local population, and before it slowly became 'self-containing'. Finally, as it was the ultimate *raison d'être* of the New Bombay project to relieve the old city of Bombay from its worst urban pressure, we have tried to assess the extent to which the new town may be having a meaningful impact on the urban situation in Bombay. Although this impact, due to the influence of many other variables, is difficult to examine, the available data indicated that there is indeed a positive, although not very strong impact on the urban situation in the old city of Bombay.

Major questions and scope for further research

The data that were collected, and the kind of work that has been done, have generated numerous new and interesting questions. The most important question that, in my opinion, has come to the surface, is what the role of the State government has been (directly, or indirectly via the NTDA) in the process of urban development in New Bombay in general, and in the shift 'up-market' in particular? To what extent has the development authority, being a fully owned company of the State government, been able to plan and develop the new town independently from government interference, and have its policies and actions been autonomous decisions? The question must be raised in order to identify who really pulls the strings in New Bombay, and why.

To gain some insight into the role of the government in the urban development processes in New Bombay it may be useful to have a look at, and briefly discuss, the same processes in the city of Bombay over the last several decades. These processes in Bombay may, to some extent, be an indication of what is currently happening in New Bombay in the field of urban development.

The major cause of many of Bombay's urban problems, it being an island with limited availability of land, is often said to be its restriction in land. This assumption I believe is wrong. The major cause of many of these problems is not the land, but rather the way in which that land is being used. Due to the existence of several old land laws, such as the Urban Land Ceiling and Regulation Act (ULCAR Act) and the Rent Control Act (RCA), the supply of land and housing in Bombay has been restricted significantly for about twenty years.[1] Due to the RCA, an estimated 20,000 flats in Bombay are currently lying vacant, and tens of thousands of flats are in a seriously dilapidated condition.[2] More importantly, although the ULCAR Act may to some extent have been a successful instrument in the hands of the State government to prevent the private sector from monopolising the urban land and from playing the role of speculator, in practice, however, it seems that the State government has largely taken over that role and is now playing the businessman at a large scale itself; the kind of role that is played by several other state governments in India, directly or indirectly, via specially established urban development agencies with almost unlimited powers and a near monopoly control over land (Misra 1986, Maitra 1991, Datta 1992). Consequently, it is not so much State legislation in itself that has caused major distortions in Bombay's urban land markets, but rather the attitude of the State government in land use and land management.

This conclusion is further exemplified by several other observations. First, while the land in South and Central Bombay is extremely intensively used, a very large area of government land (25 so-called 'no-development zones' in the northern and north-eastern suburbs, estimated to be almost 30,000 hectares in total; Sharma 1991:415) is basically lying vacant or is not properly used.[3] Due to such government-induced and unnecessary restrictions in the supply of land and housing, in combination with a rising demand, prices for land and property in Bombay have increased several times. Whereas in most western cities the 'median for the housing market' is less than 2.5, and for a large number of Asian capital cities it is between 2.5 and 5.0, for Bombay (only a regional capital city) it was, in 1995, over 14 (Lobo 1995)![4] The same counts for commercial property rates. In August 1996, prices for office space (whether sold or rented) were in Bombay the highest in the world, and were actually higher than rates in Hong Kong, Tokyo, and New York. Richard Ellis' figures of August 1996 have put the cost of office space in Bombay at US\$ 157 per ft^2 per annum, which was 27.64% higher than in Hong Kong and 37.71% higher than in Tokyo (The Asian Age, 4 August 1997). Another indicative observation is, for instance, that sales of properties acquired by the state's Income Tax Department have recently been by auction, clearly adding significantly to the rapid inflation in land and property.

The fact that there has been a negative population growth in South Bombay between 1981 and 1991, and that population growth in the Suburbs has slightly slowed down, is therefore not the result of deliberate policy. Most of these developments are mainly a result of exceptionally high property rates. Consequently, urban sprawl and population growth in Bombay continue unprecedented, only now largely beyond the limits of the area administered by the Greater Bombay Municipal Corporation, in the northern area of the Salsette Island and in the northern and eastern satellite towns in the periphery.

Some questions in relation to urban land management in Bombay therefore need an answer: (1) knowing that the government (central and state) is by far the largest landholder in Bombay, are the State government or its agents directly gaining from these policies and from the artificially high land prices, and could this be the reason why the ULCAR Act and the RCA, clearly having negative effects for the public, are not being abolished? (2) secondly, what is currently the role and function of a low so-called 'Floor Space Index' or 'Floor Area Ratio', i.e. 'the amount of floor area in a building in relation to the area of the site' (Doebele 1983:74), and what is the relation between such an FSI on the one hand, and land prices for different land use and property rates on the other? In Bombay, the FSI has always (at least officially, not considering the many corruption-related building activities) been kept as low as 1 to 1.5 all over the city, which is rather surprising given the peculiar geographical nature of the city as a peninsular and the radically different policies in other large cities with similar problems of natural land restriction such as Hong Kong or Singapore. So far, surprisingly little has been documented on this relationship between use of FSI and land prices/property rates in academic literature on urban land economics. (3) thirdly, to what extent is there a link (as is popularly believed) between State government politicians and the 'Bombay Underworld', which presently controls significant parts of the urban land and property in the city and directly gains from high land prices?

Whatever the answers to these questions and whatever the motivations behind the State government's land policies, what can be argued with certainty is that the State government, via land management and land use, plays a very important role in urban development in Bombay. That direct government interference in urban land markets can play a very significant role in Indian cities is an insight that is quite recent and has often been ignored in the past, but that has gained increasing academic support during the last decades (see for instance Amitabh 1994).

Evidently, the main question now is what the impact of such direct State government policy and activities is on the urban development process in New Bombay. Is New Bombay really planned, developed and managed autonomously by the NTDA, or is it the State government that is, to some extent, pulling the strings behind the scene? The research in this study was not really focused on this question, but has, in view of recent developments in the urban development field in New Bombay, become quite interesting.

There are certain indications that there has been a clear watershed in the State government's interest in the urban development process in New Bombay. In the first phase of development of New Bombay, roughly speaking during the 1970s and early 1980s, the State government was not interested at all in the urban development process in New Bombay. It had formally accepted this grand project in 1973, but in the same time had made it very clear that at no point it would participate financially in the project. What is most surprising, however, is that the State government never even provided the minimum of political and/or moral support to the project prior to the early 1980s, which has become clear from its policies and actions in the city of Bombay during the same period. Most of these policies and actions, indeed, went directly against the development objectives of New Bombay, and against the very *raison d'être* of the project: (a) at the same time that the NTDA was struggling to bring some development and enthusiasm to the newly developing area (with the ultimate aim to decongest South Bombay), the State government continued with extremely costly land reclamation schemes in South Bombay, and many new buildings (the new Council Hall, the new Reserve Bank of India, numerous other banks) were established here; (b) further, given the overall objectives of regional urban development, it is rather odd that two giant construction projects, the Oshiwara District Centre project and the Bandra-Kurla project, were planned in south-central Bombay to form two new major business areas.[5] (c) finally, and most significantly, the State government never showed the intention to shift even a small part of the political-administrative apparatus out of South Bombay and into New Bombay, as had been put forward in the NBDP as a major catalyst for both decongestion of Bombay and initial development of New Bombay. In short, the State government's activities in Bombay prior to the 1980s have in practice been clearly contradictory to its rhetoric to decongest the old city and to develop New Bombay as a 'counter-magnet' and major new growth pole. It was largely as a result of this attitude and political neglect that urban growth and development in New Bombay remained very sluggish during the first ten to twelve years of project implementation.

Although interviews with a number of high-ranking CIDCO-officers, among whom the Joint Managing Director, have not explicitly confirmed this viewpoint, it is my opinion that the State government, in view of the sluggish growth and problematic developments in New Bombay during the first decade, has strongly and drastically intervened and that, as a result, the policies of the development agency were radically transformed and shifted in the direction of a more commercial and privatised planning approach. Although it may well be that this particular move has been one of the major reasons why development did pick up in the period thereafter, it also implicated that a number of the earlier project objectives, such as the aim to develop a socially balanced city, did no longer seem to be of prime importance.

The argument, that the State government has been the major force behind these policy shifts, is reinforced by another observation. If the urban development process in New Bombay is not really going the way that it was planned in the NBDP, as has been demonstrated by the research, particularly in terms of social development, one could have expected the State government, being the political authority of the development agency, to have intervened. There has, however, never been any such intervention by the State government to call to account the development authority and its new policies. That the State government must have intervened significantly in the development policies and actions in New Bombay would, of course, be hardly surprising. CIDCO is a fully owned company of the State government, and its top-level officers are administrators and professionals that have been appointed by the State government. This lack of any democratic control on the development authority and its policies and actions is, of course, a major problem.

From the mid-1980s, when growth in New Bombay gradually picked up, the State government has increasingly interfered, directly or indirecly, in the urban development process in the new city. From that moment, New Bombay indeed became a very interesting area for the State government to operate in; an area in which it could practice the same policies as it has done in the old city of Bombay for the last several decades, viz. to speculate in land and to make fortunes, at the expense, unfortunately, of the ordinary New Bombay citizen. Why the State government would want to do this, against the general interest of the people, is a very large new question which may well be a research topic in itself. The nature and the role of the Indian state and the traditional power linkages in Indian society, as have been explained by such authors as Pranab Bardhan and Achin Vanaik, who's theories were formed on Poulantzas' concept of the state, may be an interesting starting point towards an answer. Entering this debate would imply to examine 'the politics of urban development', probably *the* potentially most interesting and most important field of study regarding urban development in New Bombay.

The State government has intervened in New Bombay both directly (via the same State legislation as in Bombay) and indirectly (via the policies and operations of the NTDA in New Bombay). As a result, the development authority has increasingly been showing similar characteristics and operational features as the DDA in Delhi, the earliest example in India of a state urban development agency. Interestingly, some five years ago Abhijit Datta, discussing land management in new towns in India and frequently referring to the DDA in Delhi, wrote that

> 'In effect, the development authority has assumed monopoly powers over urban land development, and its behaviour with regard to product differentiation, cross-subsidy, supply of developed land, and pricing policy have exhibited close similarities with any private monopolist without any effective regulation or accountability' (Datta, 1992:195)

CIDCO in New Bombay increasingly matches Datta's description of the DDA and other state urban development agencies in India. Over time, CIDCO has increasingly become a public development agency only in name, has been working more and more along private business lines, and is partly owned by its shareholders. New Bombay, on the other hand, has increasingly developed into a giant real estate project. These observations and conclusions are not at all surprising for those who are familiar with New Bombay and its development, either by study or profession, or by experience with living in the area. The fact that the major New Bombay advocates of the early 1970s have implicitly or explicitly confirmed these broad observations in interviews is quite significant. An important question, however, remains whether CIDCO had any other choice than to follow the State government's wishes. Being a most convenient instrument in the hands of the State government, itself making fortunes and at the same time pleasing its political allies in industrial and business circles, it may have been bound hand and foot to the State government's directives. It is an important question that remains open for speculation and further research.

To conclude, the example of New Bombay thus not only provides some very valuable positive as well as negative lessons in respect to urban planning, it also provides some potentially interesting insights into the role of the State in urban land markets and urban development in general. The very fact that it has proved to be possible to develop a new town in a developing country, combining the use of the land as a major resource with financial self-sufficiency, is a most valuable and useful observation. The major question remains, however, whether such a commercial and partly privatised planning concept as in New Bombay can, within a Third World context, go hand in hand with a more acceptable and more balanced urban development process in social terms. The New Bombay urban planning concept may carry a great potential, and may provide a number of possible solutions to the increasing problem of mega-city management in the developing world, but the social dimension of urban planning should never be ignored, as social progress and social development should be the prime concern of any planning.

Notes

[1] The ULCAR Act, which came into force in February 1976 in eleven states simultaneously (one of which was Maharashtra), limits the amount of urban land a person or company can own to 500 m². Any surplus land has to be surrendered to the state government at a minimal price, primarily to prevent that land would be concentrated in few land ownerships, and would thus, for speculative reasons, not be properly used. Furthermore, the ULCAR Act also prohibits a person to transfer

any vacant land or property without the written consent of the authorities. Thus, the ULCAR Act not only puts a ceiling on vacant urban land, it also regulates private urban property transactions. The RCA, on the other hand, basically froze rental prices at the 1947 level.

[2] The effect of the RCA on the rental market is presently such, that one family may pay not more than Rs. 500 (less than £10) rent per month, whereas the family next door, staying in a similar flat of the same size, is paying Rs. 50,000, according to the moment both households entered the rental market.

[3] A large part of this land is occupied by squatters, whereas other parts are simply lying vacant and are not even used as green zones (as was initially planned).

[4] Figure X indicates that, on an average, a person can own a house or flat by paying roughly X times his annual salary.

[5] The Oshiwara District Centre project is less known than the Bandra-Kurla project. This first project, which was originally proposed to be an alternative to Nariman Point and the Bandra-Kurla Complex (BKC), and which is valued at Rs. 1,600 crores at current real estate prices, now seems to have come to a halt. Development at the BKC, on the other hand, is in full progress. These two projects were part of the so-called *Bombay First* plan, a plan modelled after the *London First* project, aimed at making Bombay the new Singapore of western Asia, and at attracting and creating several hundred thousand jobs in banking, trade and finance.

Bibliography

1. Books - Articles – Reports

Acharya, B.P. (1987), 'The Indian Urban Land Ceiling Act. A Critique of the 1976 Legislation', *Habitat International*, vol. 11, No 3, pp. 39-51.

Angel, S., Archer, R.W., Tanphiphat, S. and Wegelin, E.A. (eds) (1983), *Land for Housing the Poor*, Select Books, Singapore.

Amitabh (1997), *Urban Land Markets and Land Price Changes. A Study in the Third World Context*, Ashgate, Aldershot, UK.

Angotti, T. (1993), *Metropolis 2000. Planning, Poverty and Politics*, Routledge New York and London.

Armstrong, W. and McGee, T.G. (1985), *Theatres of Accumulation: Studies in Asian and Latin American Urbanization*, Methuen, London and New York.

Arunachalam, B. and Deshpande, C.D. (1981), 'Bombay', in M. Pacione (ed), *Problems and Planning in Third World Cities*, Croom Helm, London.

Awasthi, R.K. (1985), *Urban Development and Metropolitics in India*, Chugh Publications, Allahabad.

Banerjee-Guha, S. (1991), 'Who are the Beneficiaries? Evaluation of a Public Housing Project for the Poor in New Bombay', *Ekistics*, vol. 58, Jan-Apr 1991.

Banerjee-Guha, S. (1989), 'Growth of a Twin City: Planned Urban Dispersal in India', in F. Costa, A. Dutt, L. Ma and A. Noble (eds), *Urbanization in Asia. Spatial Dimensions and Policy Issues*, University of Hawaii Press, Honolulu, pp. 169-188.

Banerjee-Guha, S. (1995), 'Urban Development Process in Bombay: Planning for Whom?', in A. Thorner and S. Patel (eds), *Bombay, A Metaphor for Modern India*, OUP, Bombay.

Baken, R-J. and van der Linden, J. (1992), *Land Delivery for Low Income Groups in Third World Cities*, Avebury, Aldershot, UK.

Bardhan, P. (1984), *The Political Economy of Development in India*, Basil Blackwell, Oxford UK and Cambridge, Mass. US.

Baróss, P. and van der Linden, J. (1990), *The Transformation of Land Supply Systems in Third World Cities*, Avebury, Aldershot, UK.

Batten, D. and Johansson, B. (1987), 'The Dynamics of Metropolitan Change', *Geographical Analysis*, Vol. 19, No 3, pp. 189-99.

Beier, G. (1984), 'Can Third World Cities Cope?', in P.K. Ghosh (ed), *Urban Development in the Third World*, Greenwood Press, Westport and London.

Berry (1973), *The Human Consequences of Urbanization*, Macmillan, London.

Bhardwaj, R.K. (1974), *Urban Development in India*, National Publishing House, New Delhi.

Bhattacharya, A. (1981), 'Housing for the Urban Poor: A Case Study of Bombay', in G. Bhargava (ed.), *Urban Problems and Policy Perspectives*, Abhinav Publications, New Delhi.

Bhattacharya, B. (1979), *Urban Development in India Since Pre-Historic Times*, Shree Publishing House, Delhi.

Bhattacharya, M. (1976), *Management of Urban Government in India*, Uppal Book Store, New Delhi.

Blumenfeld, H. (1971), *The Modern Metropolis*, MIT Press, Cambridge, Mass. US.

Blumer, H. (1968), *Symbolic Interactionism*, Prentice-Hall, New York.

Bogue, D.J. and Zachariah, K.C. (1962), 'Urbanisation and Migration in India', in R. Turner (ed), *India's Urban Future*, OUP, Bombay.

Bose, A. (1974), *Studies in India's Urbanisation, 1901-1971*, Studies in Demography No. 1, Tata McGraw-Hill, New Delhi.

Bose, A. (1976), *Bibliography on Urbanization in India: 1947-1976*, Tata McGraw Hill, New Delhi.

Bose, A. (1980), *India's Urbanisation 1901-2001*, Tata Mc Graw-Hill, New Delhi.

Bose, A., Desai, P.B., Mitra, A. and Sharma, J.N. (1974), *Population in India's Development 1947-2000*, Issued by the Indian Association for the Study of Population, Vikas Publishing House, Delhi.

Bradnock, R.W. (1989), *Urbanisation in India*, Case Studies in the Developing World, John Murray, London.

Bradnock, R.W. (1981), 'India's Cities: Hope for the Future?', *Geography*, vol. 66, pp. 208-220.

Brennan, E.M. and Richardson, H.W. (1989), 'Asian Mega City: Characteristics, Problems and Policies', *International Regional Science Review*, 12(2), 117-20.

Brown, L. (1984), 'The Urban Prospect: Reexamining the Basic Assumptions', in P.K. Ghosh (ed), *Urban Development in the Third World*, Greenwood Press, Westport and London.

Brunn, S.D. (1971), *Urbanization in Developing Countries: An International Bibliography*, East Lansing, Mi: Latin American Studies Center & Center for Urban Affairs, Michigan State University.

Brunn, S. and Williams, J. (eds) (1983), *Cities of the World: World Regional Urban Development*, Harper and Row, New York.

Brunn, S.D., Williams J.F. and Darden, J.T. (1983), 'World Urban Development', in S. Brunn and J. Williams (eds), *Cities of the World: World Regional Urban Development*, Harper Collins, New York.

Buch, M.N. (1988), *Planning the Indian City*, Vikas Publishing House, New Delhi. *Business India*, 11 April 1994, 'Up, Up and ... ?'

Byres, T.J. (ed) (1994), *The State and Development Planning in India*, SOAS Studies on South Asia, OUP Bombay.

Castells, M. (1977), *The Urban Question*, Edward Arnold, London.

Castells, M. (1983), *The City and the Grassroots: A Cross-Cultural Theory of Urban Social Movements*, University of California Press, Berkeley.

Chakravorty, S. (Sep 1996), 'Too Little in the Wrong Places? Mega City Programme and Efficiency and Equity in Indian Urbanisation', *Economic and Political Weekly*.

Chandrasekhar, C.S. (1974), 'India', in A. Whittich (ed), *Encyclopedia of Urban Planning*, McGraw-Hill, New York.

Chatterjee, L. and Nijkamp, P. (eds) (1983), *Urban and Regional Policy Analysis in Developing Countries*, Gower, England.

Cockburn, C. (1977), *The Local State*, Pluto, London.

Cohen, M.A. (1984), 'Cities in Developing Countries: 1975-2000', in P.K. Ghosh (ed), *Urban Development in the Third World*, Greenwood Press, Westport and London.

Correa, C.M., Mehta, M. and Patel, S.B., *Planning for Bombay. Patterns of Growth, The Twin city, Current Proposals*, so-called 'MARG Piece', n.d., prob. May 1967.

Correa, C. (1985), *The New Landscape*, The Book Society of India.

Costa, F., Dutt, A., Ma, L., and Noble, A. (eds) (1988), *Asian Urbanization. Problems and Processes*, Gebrüder Borntraeger, Berlin.

Costa, F., Dutt, A., Ma, L., and Noble, A. (eds) (1989), *Urbanization in Asia. Spatial Dimensions and Policy Issues*, University of Hawaii Press.

Costa, F., Dutt, A., Ma, L., and Noble, A. (1988), 'Divergent Paths and Policy Responses in Asian Urbanization and Planning', in F. Costa et al., *Asian Urbanization. Problems and Processes*, Gebrüder Borntraeger, Berlin.

Cotter, J.V. (1977), *New Towns as an International Phenomenon: A Bibliography*, Council of Planning Librarians, Monticello, Italy.

Crook, N. and Dyson, T. (1982), 'Urbanisation in India: Results of the 1981 Census', *Population and Development Review*, vol. 8, No. 1, pp. 145-155.

Crook, N. (1993), *India's Industrial Cities. Essays in Economy and Demography*, OUP New Delhi.

Cubukgil, A. (1981), *Urbanization in Developing Countries: Some Reflections of the Literature*, Urban and Regional Planning Department, University of Toronto.

Datta, A. (1992), 'Indian Policies on Urbanisation and Urban Development', *Third World Planning Review*, vol. 14 , No 2, 1992.

Datta, A. (1987), 'Alternative Approaches to Shelter for the Urban Poor in India', *Cities*, Feb 1987, pp. 35-42.

Davis, K. (1984), 'Asia's Cities: Problems and Options', in P.K. Ghosh (ed), *Urban Development in the Third World*, Greenwood Press, Westport and London.

Davis, K. (1962), 'Urbanisation in India: Past and Future', in R. Turner (ed), *India's Urban Future*, OUP Bombay.

Deshpande, S.L. and Deshpande, L.K. (1991), *Problems of Urbanization and Growth of Large Cities in Developing Countries: A Case Study of Bombay*, International Labour Organization, Geneva.

Doebele, W.A. (1983), 'Concepts of Urban Land Tenure', in H.B. Dunkerley (ed), *Urban Land Policy. Issues and Opportunities*, World Bank Publication, OUP New York.

Dogan, M. and Kasarda, J.D. (1987), *The Metropolis Era*, vol. II: Mega Cities, Sage Publications, California.

Drakakis-Smith, D. (1987), *The Third World City*, Methuen, London.

Drakakis-Smith, D. (ed) (1986), *Urbanisation in the Developing World*, Croom Helm, London.

Drakakis-Smith, D. (1981), *Urbanisation, Housing and the Development Process*, Croom Helm, London.

Dunkerley, H.B. (ed) (1983), *Urban Land Policy. Issues and Opportunities*, World Bank Publication, OUP New York.

Dutt, A.K. (1983), 'Cities of South Asia', in S. Brunn and J. Williams (eds), *Cities of the World: World Regional Urban Development*, Harper Collins, New York.

Edwardes, S.M. (1902), *The Rise of Bombay, a Retrospect*, Reprint of vol. X of the Census of India Series 1901, Bombay.

Erdosy, G (1988), *Urbanisation in Early Historic India*, BAR International Series 430, Oxford.

Findley, S.E. (1993), 'The Third World City: Development Policy and Issues', in J.D. Kasarda and A.M. Parnell (eds), *Third World Cities: Problems, Policies, and Prospects*, Sage Publications, California.

Gadgil, D.R. (n.d., prob. 1965), Excerpts of *Gadgil Committee's Report on Regional Plans for Bombay - Panvel and Poona Regions*.

Ghaneshwar, V. (1995), 'Urban Policies in India-Paradoxes and Predicaments', *Habitat International*, vol. 19, No. 3, pp. 293-316.

Gilbert, A. and Gugler, J. (1992, 2nd ed), *Cities, Poverty and Development. Urbanisation in the Third World*, OUP Oxford.

Gilbert, A. (1986), 'Self-Help Housing and State Intervention: Illustrated Reflections on the Petty Commodity Production Debate', in D. Drakakis-Smith (ed), *Urbanisation in the Developing World*, Routledge.

Gill, G.S., Bhattacharya, A. and Adusumilli, U. (Nov. 1995), *Sustainable Urban Development. Case Study of New Bombay, India*, CIDCO report, final draft.

Ginsberg, N. (1989), 'An Overview of the Literature', in F. Costa et al (eds), *Urbanization in Asia. Spatial Dimensions and Policy Issues*, University of Hawaii Press.

Gonsalves, C. (Aug 1981), *Bombay: A City Under Siegy*, Institute of Social Research and Education, Bombay.

Grimes, O.F. (1984), 'The Urban Housing Situation in Developing Countries', in P.K. Ghosh (ed), *Urban Development in the Third World*, Greenwood Press, Westport and London.

Gugler, J. (ed) (1988, reprint 1992), *The Urbanisation of the Third World*, OUP New York.

Gugler, J. (ed) (1996), *The Urban Transformation of the Developing World*, OUP New York.

Gupta, D. (1982), *Nativism in a Metropolis: The Shiv Sena in Bombay*, Manohar Publications, New Delhi.

Gupta, G.R. (ed), *Urban India*, Vikas Publishing House, New Delhi.

Gupta, L.C. (1985-a), 'Migration and Population Growth', in R. Mayur and P.R. Vohra (eds) (1985), *Bombay by 2000 AD*, Collection of small articles, publisher unknown.

Gupta, L.C. (1985-b), *Creative Urban Development. The CIDCO Experience in Planning, Land Assembly, Financing and Implementation*, published by CIDCO, Bombay.

Gupta, L.C. (n.d., prob. 1986), *Housing for Low Income Groups. A New Bombay Experience*, CIDCO Document.

Harris, J. (1990), *Urbanisation, Economic Development and Policy in Developing Countries*, Working Paper No. 19, Development Planning Unit, University College London.

Harris, N. (1978), *Economic Development, Cities and Planning : The Case of Bombay*, OUP Bombay.

Harris, N. (1990), 'Urbanisation, Economic Development and Policy in Developing Countries', *Habitat International*, No. 4, pp. 3-42.

Harris, N. (ed) (1992), *Cities in the 1990s. The Challenge for Developing Countries*, UCL Press, Development Planning Unit, University College London.

Harris, N. (1992), *Bombay in a Global Economy: Structural Adjustment and the Role of Cities*, Working paper, Bombay.

Harris, N. (1993), 'Land-Use in Bombay', *Economic and Political Weekly*, 16 Oct 1993, pp. 2300.

Harvey, J. (1992, third edition), *Urban Land Economics*, Macmillan, UK.

Hayes, E.C. (1972), *Power Structure and Urban Policy: Who Rules in Oakland*, McGraw-Hill, New York.

Horvath, R.J. (1969), 'In Search of a Theory of Urbanisation: Notes on the Colonial City', *The East Lakes Geographer*, vol 3 (Dec), pp. 69-82.

Hoselitz, B.F. (1962), 'Survey of the Literature on Urbanisation in India', in R. Turner (ed), *India's Urban Future*, OUP Bombay.

India Today, 25 Aug 1997, 'Murder in Mumbai'.

India Today, 6 Oct 1997, 'Fear in the City'.

India Today, 26 Jan 1998, 'The Indus Riddle'.

Instituto Del Tercer Mundo (1997), *The World Guide 1997/98*, New Internationalist Publications, Oxford.

Internationale Samenwerking (1996), 'Hoe houden we de stad leefbaar. Steden, de vieze motor van de wereldeconomie', *Internationale Samenwerking*, May '96.

Jacquemin, A. (1997), 'Tweederangsburgers moeten wijken voor tweelingstad', *De Wereld Morgen*, vol. 4.

Jain, A.K. (1990), *The Making of a Metropolis. Planning and Growth of Delhi,* National Book Organisation, New Delhi.

Jain, M.K. (1977), *Interstate Variations in the Trends of Urbanization in India,* International Institute of Population Studies, Bombay.

Kasarda, J.D. and Parnell, A.M. (eds) (1993), *Third World Cities. Problems, Policies and Prospects,* Sage Publications.

Kopardekar, H.D. (1986), *Social Aspects of Urban Development. A Case Study of the Pattern of Urban Development in the Developing Countries,* Popular Prakashan, Bombay.

Kosambi, M. (1986), *Bombay in Transition : The Growth and Social Ecology of a Colonial City, 1880-1980,* Almqvist en Wiksell International, Stockholm.

Kundu, A. (July 1983), 'Theories of City-Size Distribution and Indian Urban Structure: A Reappraisal', *Economic and Political Weekly.*

Kundu, A. (May 1989), 'National Commission on Urbanisation: Issues and Non-Issues', *Economic and Political Weekly.*

Kundu, A. and Sharma, R.K. (1984), 'Industrialisation, Urbanisation and Economic Development', *Urban India,* Vol. 43, No 1, pp. 43-50.

Lea, J.P. and Courtney, J.M. (eds) (1985), *Cities in Conflict: Studies in the Planning and Management of Asian Cities,* World Bank Symposium, Washington.

Linn, J.F. (1983), *Cities in the Developing World: Policies For Their Equitable and Efficient Growth,* Oxford University Press.

Lipton, M. (1988), 'Why Poor People Stay Poor: Urban Bias in World Development', in J. Gugler (ed), *The Urbanisation of the Third World,* OUP New York.

Lojkine, J. (1976), 'Contribution to a Marxist Theory of Capitalist Urbanisation', in C.G. Pickvance (ed), *Urban Sociology: Critical Essays,* Tavistock London.

Maitra, Sipra (1991), 'Housing in Delhi. DDA's Controversial Role', *Economic and Political Weekly,* 16 Feb 1991, pp. 344-46.

Mallick, U.C. (1981), 'Urban Land (Ceiling and Regulation) Act: An Appraisal for Effective Solution of Urban Problems', in G. Bhargava (ed), *Urban Problems and Policy Perspectives,* Abhinav Publications, New Delhi.

Mandal, R.B. (1982), *Urbanization and Regional Development,* Concept Publishing House, New Delhi.

Mathur, O.P. (1994), 'Responding to the Urban Challenge: A Research Agenda for India and Nepal', in R.L. Stren (eds), *Urban Research in the Developing World,* Vol. I, Toronto, 189-99.

Mayur, R. and Vohra, P.R. (eds) (1986), *Bombay by 2000 AD,* Collection of articles.

McGee, T.G. (1995), 'Eurocentricism and Geography: Reflections on Asian Urbanisation', in J. Crush (ed), *Power of Development,* Routledge, London, pp. 192-210.

McGee, T.G. (1967), *The Southeast Asian city: a social geography of the primate cities of Southeast Asia*, G. Bell London.

Meier, R.L. (Feb 1972), *Resource Conserving Urbanism in South Asia IV: The Development of Greater Bombay*, Working Paper No. 154, Institute of Urban and Regional Development, University of California, Berkeley.

Mills, E. and Becker, C. (1986), *Studies in Indian Urban Development*, World Bank Research Publication, OUP New York.

Misra, B. (1986), 'Public Intervention and Urban Land Management. The Experience of Three Metro-Cities of India', *Habitat International*, vol. 10, No 1/2, pp. 59-77.

Misra, R.P. (1978), 'Towards a Perspective Urbanisation Policy', in R.P. Misra, (ed), *Million Cities in India*, Vikas Publishing House, Delhi.

Mohan, R. (1985), 'Urbanization in India's Future', *Population and Development Review*, 11, pp. 619-45.

Mohan, R. (1996), 'Urbanization in India: Patterns and Emerging Policy Issues', in J. Gugler (ed), *The Urban Transformation of the Developing World*, OUP New York.

Mullick, U.C. (1981), 'Urban Land (Ceiling and Regulation) Act: An Appraisal for Effective Solution of Urban Problems', in G. Bhargava (ed), *Urban Problems and Policy Perspectives*, Abhinav Publications, New Delhi.

Murphey, R. (1996), 'A History of the City in Monsoon Asia', in J. Gugler (ed), *The Urban Transformation of the Developing World*, OUP New York.

Nath, V. (1989), 'Urbanization in India: Retrospect and Prospect', in F. Costa et al, *Urbanization in Asia. Spatial Dimensions and Policy Issues*, University of Hawaii Press, Honolulu.

Newell, M. (1977), *An Introduction to the Economics of Urban Land Use*, The Estates Gazette Ltd., London.

Nientied, P. and van der Linden, J. (1990), 'The Role of the Government in the Supply of Legal and Illegal Land in Karachi', in P. Baróss and J. van der Linden, *The Transformation of Land Supply Systems in Third World Cities*, Avebury UK.

Nientied, P. and van der Linden, J. (1988), 'Approaches to Low-Income Housing in the Third World', in J. Gugler (ed), *The Urbanisation of the Third World*, OUP New York.

NIUA (1983), *Integrated Urban Development Programme for Metropolitan Cities and Areas of National Importance. An Evaluation*, National Institute of Urban Affairs, New Delhi.

NIUA (1988), *State of India's Urbanisation*, National Institute of Urban Affairs, New Delhi.

NIUA (May 1994), *Privatization of Land Development and Urban Services: A Case Study of CIDCO*, National Institute of Urban Affairs, New Delhi.

Noble, A.G. and Dutt, A.K. (eds) (1977), *Indian Urbanization and Planning: Vehicles of Modernisation*, Tata Mc Graw-Hill, New Delhi.

Operations Research Group (April 1993), *Study of Municipal Services Requirements in the Proposed Municipal Corporation Areas in New Bombay*, Report submitted to CIDCO by ORG, Baroda.

Pacione, M. (ed) (1981), *Problems and Planning in Third World Cities*, Croom Helm, London.

Patel, S.B. (1993), 'A Second Financial Centre for Bombay. Where Should It Be?', *Economic and Political Weekly*, 7 Aug 1993, pp. 1631.

Payne, G.K (1977), *Urban Housing in the Third World*, Leonard Hill, London and Routledge & Kegan Paul, Boston.

Pickvance, C. (1995), 'Marxist Theories of Urban Politics', in Judge, D., Stoker, G. and Wolman, H. (eds), *Theories of Urban Politics*, Sage Publications London, California and New Delhi.

Potter, Robert (1992, repr. 1993), *Urbanisation in the Third World*, Oxford University Press.

Prakasa R.V.L.S. (1983), *Urbanisation in India: Spatial Dimensions*, Concept Publishing House, New Delhi.

Premi, M.K. (1991), 'India's Urban Scene and its Future Implications', *Demography India*, Vol. 20, No 1, pp. 41-52.

Premi, M.K. and Tom, J.A.L. (1985), *City Characteristics, Migration and Urban Development Policies in India*, East-West Population Institute No 92., E-W Population Institute, Honolulu.

Preston, S.H. (1988), 'Urban Growth in Developing Countries: A Demographic Reappraisal', in J. Gugler (ed), *The Urbanisation of the Third World*, OUP New York.

Prins, W.J.M. (1994), *Urban Growth and Housing Delivery, Past and Present. A Comparative Analysis of Nineteenth-Century London and Contemporary Delhi*, Ph.D. thesis University of Leiden, Leiden Development Studies, No. 14.

Rajagopalan, C. (1962), *The Greater Bombay. A Study in Suburban Ecology*, Popular Book Depot, Bombay.

Ramachandran, R. (1989), *Urbanization and Urban Systems in India*, OUP Delhi.

Ramachandran, P. (1977), *Housing Situation in Geater Bombay*, Tata Institute of Social Sciences Series No. 38, Somaiya Publications Ltd. Bombay/New Delhi.

Rao, G.B.K. (1981), 'Urban Land (Ceiling and Regulation) Act (1976): Will it Fulfil Goals of Urban Development?', in G. Bhargava (ed), *Urban Problems and Policy Perspectives* (Abhinav Publications New Delhi).

Renaud, B. (1981), *National Urbanization Policy in Developing Countries*, World Bank Report, OUP.

Richardson, H.W. (1984), 'National Urban Development Strategies (NUDS) in Developing Countries', in P.K. Ghosh (ed), *Urban Development in the Third World*, Greenwood Press, Westport and London.

Richardson, H.W. (1987), 'Whither National Urban Policy in Developing Countries?', *Urban Studies*, Vol. 24, pp. 227.

Roberts, B. (1978), *Cities of Peasants. The Political Economy of Urbanisation in the Third World*, Sage Beverly Hills.

Rondinelli, D.A. (1984), 'Balanced Urbanization, Regional Integration and Development Planning in Asia', in P.K. Ghosh (ed), *Urban Development in the Third World*, Greenwood Press, Westport and London.

Sabade, B.R. (ed) (1987), *Industrial Development of Maharashtra*, Mahratta Chamber of Commerce and Industries, Pune.

Sealey, N. (1988), 'Planned Cities of India: A Study of Jaipur, New Delhi and Chandigargh', in F. Costa et al, *Asian Urbanization. Problems and Processes*, Gebrüder Borntraeger, Berlin.

Sharma, P. (1995), 'Future Metros', *India Today*, 30 Sep 1995, pp. 128-41.

Sharma, R.N (1991), 'Land Grab, Bombay Style. Urban Development in Vasai-Virar Hinterland of Bombay', *Economic and Political Weekly*, 23 Feb 1991, pp. 413-17.

Sharpe, W.R.S. (1930), *Bombay, the gateway of India*, Bombay Port Trust, Bombay.

Shaw, A. (1994), 'Urban Growth and Land Use Conflicts: The Case of New Bombay', *Bulletin of Concerned Asian Scholars*, Vol. 26, No 3, Colorado.

Shaw, A. (1995), 'Satellite Town Development in Asia: The Case of New Bombay', *Urban Geography*, Vol. 16, No 3, pp. 254-71.

Sivaramakrishnan, K.C. and Green, L. (1986), *Metropolitan Management: The Asian Experience*, OUP World Bank, New York.

Sjoberg, G. (1960), *The Preindustrial City, Past and Present*, The Free Press, Glencoe, IL, US.

Slater, D. (1986), 'Capitalism and Urbanisation at the Periphery: Problems of Interpretation and Analysis with Reference to Latin America', in D. Drakakis-Smith (ed) (1986), *Urbanisation in the Developing World*, Routledge London.

Statistical Outline of India 1994-95, Tata Services Ltd. Department of Economics and Statistics, Bombay.

Stren, R. (ed) (1994-95), *Urban Research in the Developing World*, four volumes, Centre for Urban and Community Studies, University of Toronto.

Sundaram, K.V. (1977), *Urban and Regional Planning in India*, Vikas Publishing House, Delhi.

Sundaram, P.S.A. (1989), *Bombay, Can It House its Millions?*, Clarion Books New Delhi.

Taylor, J.L. and William, D.G. (1982), *Urban Planning Practice in Developing Countries*, Pergamon Press, Oxford and New York.

Todaro, M. (1984), 'Urbanization in Developing Nations: Trends, Prospects and Policies', in P.K. Ghosh (ed), *Urban Development in the Third World*, Greenwood Press, Westport and London.

Tolley, G.S. and Vinod, T. (1987), *The Economics of Urbanization and Urban Policies in Developing Countries*, World Bank Symposium, Part II, Washington DC.

United Nations (1976), *Global Review of Human Settlements*, UN Department of Economic and Social Affairs, Pergamon Press, New York.

United Nations (1978), *Concise Report on Monitoring of Population Policies*, UN Economic and Social Council, Population Commission, New York.

United Nations (1991), *Cities, People and Poverty: Urban Development Co-operation for the 1990s*, UNDP Strategy Paper, New York.

United Nations (1994), *Prospects of World Urbanisation 1994*, Department of International Economic and Social Affairs, New York.

United Nations (1996), *The State of World Population 1996*, UN Population Fund, New York.

United Nations (1997), *India. Towards Population and Development Goals*, UN Population Fund for UN System in India, OUP.

Vanaik, A. (1990), *The Painful Transition. Bourgeois Democracy in India*, Verso, London and New York.

Verma, H.S. (1981), 'Land as a Resource for Developing a New City: Rhetoric, Operationalisation, and Lessons from New Bombay', *Nagarlok*, vol. xiii, July-Sep 1981, No.3, pp. 29-65.

Verma, H.S. (1985), *Bombay, New Bombay and Metropolitan Region. Growth Process and Planning Lessons*, Concept Publishing Company, New Delhi.

Verma, N. (1981), 'Urban Land Acquisition: Certain Emerging Issues', in G. Bhargava (ed), *Urban Problems and Policy Perspectives*, Abhinav Publications, New Delhi.

Wadhva, C.D. (1980), 'Substitution of Octroi', in A. Datta (ed), *Municipal and Urban India*, Selections from 'Nagarlok', Indian Institute of Public Administration, New Delhi.

Wadhva, K. (1991), 'Delhi Rent Control Act: Facts and Fallacies', *Economic and Political Weekly*, 25 May 1991, pp. 1351-56.

World Bank (1972), *Urbanization*, Working paper, Washington DC, June 1972.

World Bank (1991), *Urban Policy and Economic Development: An Agenda for the 1990s*, World Bank Policy Paper, Washington DC.

Yadav, C.S. (1987), *City Planning, Problems and Prospects*, Perspectives in Urban Geography, vol. 14, Concept Publishing Company, New Delhi.

Yadava, K.N.S. (1989), *Rural-Urban Migration in India. Determinants, Patterns and Consequences*, Independent Publishing Company, New Delhi.

2. Government Material (published and unpublished)

BDAC (Aug 1987), *Report on Bombay's Second Development Plan 1981-2001*, Bombay Development Advisory Committee.

Bhattacharya, A. (June 1971), *A Study of Industries in Trans-Thana Belt*, CIDCO.

BMRDA (not dated, b/w 1981-87), *Shaping Bombay of the Ninetees*, Bombay Metropolitan Region Development Authority, Bombay.

BMRDA (March 1991), *Basic Transport and Communication Statistics for BMR,* Transport and Communication Division, BMRDA.

BMRDA (1992), *Bombay Metropolitan Region Development Authority. A Review.,* BMRDA, Bombay.

BMRDA (Oct 1995), *Draft Regional Plan for Bombay Metropolitan Region 1996-2011,* BMRDA, Bombay.

BMRDA (1996), *Mega City Scheme for Bombay,* BMRDA, Bombay.

BMRPB (March 1974), *Regional Plan for Bombay Metropolitan Region 1970-91,* BMRPB, Bombay.

CIDCO (May 1970), *Memorandum of Association and Articles of Association.*

CIDCO (1972), *Project Report for Vashi 1972-73.*

CIDCO (Aug 1972), *Early Development Area Vashi New Bombay. Project Report.*

CIDCO (Oct 1973), *New Bombay Draft Development Plan.*

CIDCO (1978-1996), Annual Reports.

CIDCO (Oct. 1979), *Project Report for Vashi Node 1979-80.*

CIDCO (1981), *Multi Sector Urban Development Projects for New Bombay.*

CIDCO (May 1983), *An Outline of Activities.*

CIDCO (1983), *Project Report for Vashi Node 1983-84.*

CIDCO (Aug 1984), *A Report on Survey of Industries in Thana-Belapur Area in New Bombay.*

CIDCO (Nov 1986), *New Bombay, An Outline of Progress.*

CIDCO (1988), *A Report on Socio-Economic Survey of Households in various Nodes in New Bombay.*

CIDCO (July 1989), *New Bombay, An Outline of Progress.*

CIDCO (1990), *A Report on the Survey of Industries in New Bombay.*

CIDCO (May 1992), *Two Decades of Planning & Development.*

CIDCO (1995/early 1996), *Illustrations of Good Practices Followed by CIDCO in Development of New Bombay* (internal document).

CIDCO (April 1995), *An Action Plan for Shelter Sector in New Bombay* (internal document on new housing policy).

CIDCO (May 1995), *Revised Project Report of Vashi-Sanpada Node in New Bombay.*

CIDCO (Feb 1996), *Capital Expenditure and Receipts of CIDCO.*

CIDCO (not dated), *A Brief Note on CIDCO's Land Pricing Policy and Method of Fixing Reserve Price of Land* (internal document).

Government of India (1975), *Census of India 1971, General Population Tables, Series 1.,* Government of India Press, New Delhi.

Government of India (Aug 1988), *Report of the National Commission on Urbanisation,* Vol. II.

Government of India (1991), *Census of India 1991, Provisional Population Totals, Series 1,* Government of India Press, New Delhi.

Government of India, *Census data for Maharashtra 1881-1991.*

Government of Maharashtra, Census of India 1971, District Census Handbook *Greater Bombay* (Hindi version).

Government of Maharashtra (March 1974), *Regional Plan for Bombay Metropolitan Region 1970-1991*, Town Planning and Valuation Department, Government of Maharashtra, Poona.

Government of Maharashtra, *Census of India 1981, District Census Handbook, Greater Bombay*, Maharashtra Census Directorate Bombay.

Government of Maharashtra, *Census of India 1991, Series 14, Maharashtra*, Directorate of Census Operations, Maharashtra.

Government of Maharashtra, Directorate of Industries (June 1995), *Investment and Job Creation in Maharashtra since July 1991.*

Government of Maharashtra (Public Works Department), *Basic Transport Statistics 1992-95.*

Government of Maharashtra (Nov. 1987), *Selected Indicators for Districts in Maharashtra and States in India, 1984-85.*

Government of Maharashtra (1989), *Industrial Maharashtra. Facts, Figures and Opportunities*, Maharashtra Economic Development Council.

Government of Maharashtra (Aug. 1990), *Selected Indicators for Districts in Maharashtra and States in India, 1987-88.*

Government of Maharashtra (Aug. 1993), *Selected Indicators for Districts in Maharashtra and States in India 1990-1991.*

Government of Maharashtra (1984), *Report of the Fact Finding Committee on Regional Imbalance in Maharashtra*, Government of Maharashtra Planning Department, Bombay.

Maharashtra State Gazetteers 1964-Kolaba District, Directorate of Printing and Stationary, Maharashtra State, Bombay.

MCGB (1995), *Socio-Economic Review of Greater Bombay (1993-94)*, Centre for Research & Development, Municipal Corporation of Greater Bombay.

3. Research Methodology

Abbot, J. (not dated, 1995 or 1996), *PRA and Research: How Can They Be Integrated?*, paper included in the IDS PRA Topic Pack, Institute of Development Studies, University of Sussex, UK.

Andranovich, G.D. and Riposa, G. (1993),*Doing Urban Reserach*, Applied Social Research Methods Series No. 33, Sage Publications, London.

Atkinson, P. and Coffey, A (1996), *Making Sense of Qualitative Data. Complementary Strategies*, Sage Publications, London.

Attwood, H., Gaventa, J. and Cornwall, A. (1997), *Participatory Research: Ideas on the Use of Participatory Approaches by Post Graduate Students and Others in Formal Learning and Research Institutes*, IDS PRA Topic Pack, Institute of Development Studies, University of Sussex, UK.

Burgess, R.G. (ed) (1980), *Field Research: A Sourcebook and Field Manual*, Allen & Unwin, London.

Burgess, R.G. (ed) (1984), *In The Field: An Introduction to Field Research*, Allen & Unwin, London.

Campbell, G.J., Shrestha, R. and Stone, L. (1979), *The Use and Misuse of Social Science Research in Nepal*, Research Centre for Nepal and Asian Studies, Tribhuvan University, Kirtipur, Kathmandu.

Cassel, C. and Symon, G. (eds) (1995), *Qualitative Methods in Organizational Research. A Practical Guide*, Sage Publications.

Chambers, R. (Oct 1992), *Rural Appraisal: Rapid, Relaxed and Participatory*, Discussion Paper 311, Institute of Development Studies, University of Sussex, UK.

Coffey, A. and Atkinson, P. (1996), *Making Sense of Qualitative Data*, Sage Publications.

Denzin, N. and Lincoln, Y. (eds) (1994), *Handbook of Qualitative Research*, Sage Publications, London.

Fowler Jr. and Floyd J. (1993, 2nd ed), *Survey Reasearch Methods*, Sage Publications.

Franzel, S. and Crawford, E. (1987), 'Comparing Formal and Informal Survey Techniques for Farming Systems Research: A Case Study from Kenya', *Agricultural Administration*, No. 27, pp. 13-33.

Galtung, J. (1967), *Theory and Methods of Social Research*, Allen and Unwin, London.

Gill, G.J. (1992), *But How Does it Compare to the Real Data? Lessons from an RRA Training Exercise in Western Nepal*, HMG Ministry of Agriculture-Winrock International, Kathmandu, Nepal.

Hammersley, M. and Atkinson, P. (1983), *Ethnography: Principles in Practice*, Tavistock, London.

Inglis, A. (1990), *Harvesting Local Forestry Knowledge. A Field Test and Evaluation of RRA Techniques for Social Forestry Project Analysis*, MSc. Dissertation, University of Edinburgh.

Kvale, S. (1996), *Interviews. An Introduction to Qualitative Research Interviewing*, Sage Publications, London.

Mason, J. (1997), *Qualitative Researching*, Sage Publications.

Miles, M.B. and Huberman, M.A. (1994), *Qualitative Data Analysis. An Expanded Sourcebook*, Sage Publications.

Mukherjee, N. (1995), *Participatory Rural Appraisal and Questionnaire Survey. Comparative Field Experience and Methodological Innovations*, Concept Publishing Company, New Delhi.

Norrish, P. (1996), some unpublished remarks and comments on the use of PRA in academic research (correspondance between Norrish and Attwood in the course of the organisation of a one-day workshop on the use of PRA in November 1996 at the IDS, University of Sussex, UK).

PLA Notes (1988-1997), *Sustainable Agriculture Programme*, International Institute for Environment and Development, London. (formerly *RRA Notes*)

Pretty, J.N. (1995), 'Participatory Learning for Sustainable Agriculture', *World Development*, vol. 23, number 8, August 1995, pp. 1247-1263.

Rubin, H. and Rubin, I. (1996), *Qualitative Interviewing. The Art of Hearing Datal*, Sage Publications, London.

Scoones, I. (1995), 'Investigating Difference: Applications of Wealth Ranking and Household Survey Approaches among Farming Households in Southern Zimbabwe', *Development and Change*, vol. 26, 1995, pp. 67-88.

Silverman, D. (1993), *Interpreting Qualitative Data. Methods for Analysing Talk, Text and Interaction*, Sage Publications, London.

Stenhouse, L (1981), 'Using Case Study in Library Research', *Social Science Information Studies*, vol. 1, No. 4, pp. 221-230.

Wolcott, H.F. (1994), *Transforming Qualitative Data. Description, Analysis, and Interpretation*, Sage Publications, London.

4. Newspaper Articles - Unpublished Material

Amitabh (1994), *Residential Land Price Changes in Selected Peripheral Colonies of Lucknow City, India, 1970-1990*, Unpublished Ph.D. Dissertation, University of Cambridge.

Bhattacharya, A. (1995), *An Integrated Approach to Urban Development. A CIDCO Experience in New Bombay*, paper written for publication with the All India Housing Development Association, AIHDA.

Blomkvist, H. (1988), *The Soft State: Housing Reform and State Capacity in Urban India*, Unpublished PhD thesis, Uppsala University.

Deshmukh, B.G , 'Is Mumbai Chicago?', *The Asian Age*, 4 Sep 1997.

D'Souza J.B. (1986-a), 'Hovels are not Havens', *The Times of India*, 22 May 1986.

D'Souza J.B. (1986-b), 'Ceilings without Homes', *The Times of India*, 17 August 1986.

D'Souza J.B. (1986-c), 'Bombay's Transport System: Extravagance is Easier', *The Times of India*, 9 November 1986.

Fernandes, A. (1994), 'The Pressures of Over-Urbanisation are taking many of India's Metros to the Brink', *The Sunday Times of India*, 28 August 1994.

Financial Express, 7 March 1995, 'Gold From Marches'.

Financial Express, 7 March 1995, 'A Crunch for Space'.

Financial Express, 14 March 1995, 'Up, Up and Away'.

Financial Times, 19 June 1995, 10-page FT-survey on Maharashtra.

Hafeez, P., 'The ghost of Mafia cannot be wished away', *The Asian Age*, 27 Aug. 1997.

Honkalse, S. (1992), *The Evolution and Planning of New Towns: A Study of New Bombay as a Counter Magnet to Bombay City*, unpublished M.Phil. dissertation, Department of Economics, University of Bombay, January 1992.

Jacquemin, A. (1997), *The Politics of Urban Development in New Bombay*, Paper presented at the 'Fifth Asian Urbanisation Conference', London, 26-30 Aug. 1997.

Jacquemin, A. (1997), *Bijzondere processen van politiek geïnduceerde sociale mobiliteit in Maharashtra, India*, paper presented at the conference of the Netherlands Association for Asia and Pacific Studies (NVAPS), Utrecht, 4 Oct. 1997.

Jacquemin, A. (1998), *Urban Planning in New Bombay. Physical and Socio-Economic Growth and Development of a 'Counter-Magnet' in India*, PhD. dissertation, School of Oriental and African Studies, University of London.

Kashekar, P.B. (Dec. 1985), *Note on City and Industrial Development Corporation of Maharashtra.*

Kashelikar, S. (1995), *Mobilisation of Resources Through Property Tax Administration*, Unpublished Ph.D. thesis SNDT University Bombay.

Kumar, S. (1995), 'New Bombay stations spruced up at commuter's cost', *Indian Express*, 8 Jan 1995.

Lobo, F. (1995), 'Clone City or Oasis of Peace?', *Metropolitan on Saturday*, 8 April 1995.

Mackie, J. (n.d.-early 1980s), *Regional Planning and Industrial Location Policy in India*, Discussion paper, Department of Geography, SOAS.

Mukherji, S (1994), 'Countdown to 2010. Urban atrophy', *The Independent*, 22 July 1994.

National Housing Bank (June 1992), *Report on Trend and Progress of Housing in India*, Unpublished paper.

Panjwani, N. (1994), 'System Overload', *The Independent*, 22 July 1994.

Pasricha, P.S. (1994), 'Pathway to Privatisation', *The Sunday Free Press*, 13 November 1994.

Patel, S.B. (1996), *Urban Slums*, Paper prepared for publication in The Hindu.

Prasad, R. and Gupta, K. (1995), *Patterns and Causes of Migration to New Bombay*, Unpublished Research Report from the International Institute for Population Sciences, Bombay.

Rameshkumar, M. (1994), *Enabling the Urban Poor to Grow with the New City - The Case of New Bombay*, Unpublished report of IHS Rotterdam, July 1994.

Ray, M (1989), *Study of Vashi: A Residential Node of New Bombay*, unpublished MPhil. dissertation, Department of geography, University of Bombay.

Report of the Study Group on Greater Bombay (1959) (so-called 'Barve Report'), Publisher unknown.

Sridhar, R., 'Why The Price Rise?', *The Times of India*, 29 Oct 1994.

The Accommodation Times, 1986-97 issues.

The Asian Age, 10 Jan 1995, 'Housing Prices in Bombay Going Up by 70% Every Year'.

The Asian Age, 6 March 1995, 'Bombay real estate prices boom as NRIs, MNCs step in'.

The Asian Age, 15 April 1995, '48% growth rate achieved by JNPT'.

The Asian Age (editorial), 23 Aug 1997, 'Underworld on top'.

The Asian Age, 4 Aug 1997, 'Mumbai is costliest city to work in'.

The Indian Express (Navi Mumbai supplement), 3 May 1997, 'Bill for Resettlement'.

Tinaikar, S.S. (1995), 'The Lost City', *Mid-Day*, 24 Jan 1995.

Tinaikar, S.S. (1994), 'The Fall of Bombay', *Mid-Day*, 22 Dec 1994.

Tinaikar, S.S. (1995), 'Uncontrolled Urban Sprawl', *Mid-Day*, 5 Jan 1995.

Appendices

Appendix 1: Methodology of the village survey

In chapter seven we have discussed the results of the village survey, which was based on the research techniques of Rapid Rural Appraisal (RRA) and to a lesser extent on Participatory Rural Appraisal (PRA). Both RRA and PRA are based on interviewing, which is one of the possible research methods that are available in quantitative and qualitative social science research. In order to clarify the place and functions of RRA (and PRA), we will in this Appendix briefly discuss the major distinction between quantitative and qualitative research, move on to have a quick look at the research method of interviewing in general, to further narrow it down and to come to a brief discussion of the concept, the choice, the practical usefulness and the accuracy of RRA and PRA research.

A) Positivist-quantitative vs. interpretative-qualitative research

One of the major divides in contemporary scientific social research is the discussion that has been going on between the positivist 'school' of social science, which seeks to test correlations between variables and mainly uses quantitative research techniques and methods, and the interpretative 'school', which is more concerned with observation and description and mainly uses qualitative research techniques. In quantitatively oriented texts, qualitative research is often treated as a relatively minor methodology, to be used (if at all) at early stages of a study to familiarise oneself with a setting before the serious quantitative sampling and counting begins. In this view, even though some quantitative researchers approve of the use of qualitative research when one knows relatively little about the subject under investigation, statistical analysis is generally assumed as the basis of research.

Although these schools have, unfortunately, sometimes been defined as polar opposites, it must be clear that both, quantitative and qualitative research, have their own merits and value. The choice of one or the other must depend, not on some old-fashioned prejudice, but exclusively on a number of basic given elements, such as the field of research, the aims of the research, the location, the respondents, possible limits such as cultural habits and attitudes, or available budget and time frame. The most important element, however, must be the *kind* of information that is being sought. It thus all depends on what you are trying to do, and what you are trying to prove. It is evident, that quantitative research will not always come out as the most appropriate research technique independently of what the context or situation is. C. Wright Mills, already in 1953, pointed out that a lot of quantitative research can be considered as 'abstracted empiricism'. Some time later, Blumer for example noted how attempts to establish correlations between variables depends upon a lack of attention to how these variables are defined by the people being studied. On the other hand, Socrates, in Ancient Greece, encouraged understanding by asking his students pointed questions; an early form of the open-ended informal interviews in contemporary research. In modern times, Silverman (1993) points out that in British market research circles, qualitative research is the latest fashion, and is seen to provide 'in-depth' material which is believed to be absent from quantitative survey research data (Silverman:23).

What exactly qualitative research is, is difficult to describe, as there is no standard approach among qualitative researchers. First, a lot of qualitative research or field research has to do with 'common sense'. Silverman argues that one of the traps of the 'absolutist school' is exactly the temptation to automatically accept such conventional wisdoms of our day, such as, for example, *scientism*, which involves uncritically accepting that 'science' is both highly distinct from, and superior to such 'common sense'. Second, qualitative research has some distinct features, for example that it can provide a broader version of theory than simply a relationship between variables. Another important characteristic of qualitative research is that it is very flexible, and that it allows theory development to be pursued in a highly effective and economical manner. Moreover, qualitative research, by now, has had sufficient time to develop and to build cumulative bodies of knowledge.

Social science research distinguishes four major research methods, which may be used in one combination or another: observation, analysing texts and documents, interviewing, and recordings and transcripts. These research methods are being used by both quantitative and qualitative researchers, but with different goals and in different ways. For interviewing, for example, quantitative researchers will usually prefer 'fixed-choice questions' (*yes* or *no*) because the answers they produce lend themselves to simple tabulation. Qualitative researchers, on the other hand, will often choose 'open ended questions' which produce answers which need to be subsequently coded. With the latter, authenticity rather than reliability is often the issue.

This is not to argue, of course, that reliability, validity and accuracy are of no or less importance in qualitative research, nor that they are per definition weak. Although some would go so far as to say that in qualitative research 'anything goes', these kind of assumptions must be rejected, as questions of validity and reliability are of uttermost importance in both qualitative *and* quantitative research. For what the former is concerned, we simply cannot be satisfied merely with what Silverman occasionally describes as 'telling convincing stories'.

B) Interviewing as a research method

Since Rapid Rural Appraisal is a survey technique belonging to the methodological family of interviewing, we must briefly look into this research method, which Burgess has described as 'conversations with a purpose' (Burgess 1984:102). There are two distinct ways to proceed with the interviewing technique. The conventional, positivist approach of quantitative research mainly seeks to generate interview data which are valid and reliable, independently of the research setting.[1] The main ways to achieve this are the random selection of the interview sample and the administration of standardised questions with multiple-choice answers which can be readily tabulated. In contrast, the 'interactionist approach' of qualitative research, considers the respondents as experiencing subjects who actively construct their social worlds, and the primary issue here is to generate data which give an authentic insight into people's experiences. The main ways to achieve this are unstructured, open-ended interviews, usually based upon prior, in-depth participant observation (Silverman, 1993:90-114). For positivists, the interviewer-respondent interaction is strictly defined by the research protocol, whereas for interactionists the interviews are essentially about symbolic interaction, and are more like social events. Or as Hammersley and Atkinson (1983) put it, 'Interviews must be viewed ... as social events in which the interviewer is a participant observer' (p. 126).

So far as the reliability of interview data is concerned, Burgess for example sees the interview as giving greater depth than other research techniques, because it is based on a sustained relationship between the informant and the researcher (Burgess, 1980:109). This is why most interactionists tend to reject pre-scheduled standardised interviews and prefer open-ended interviews, as they allow respondents to use their unique ways of defining the world and to raise important issues that are not contained in the schedule, and as they assume that no fixed sequence of questions is suitable to all respondents. Therefore, as Silverman concludes, 'interviews offer a rich source of data which provide access to how people account for both their troubles and good fortune' (Silverman, 1993:114).

C) Choice, use and validity of RRA research

The philosophy, approaches and methods known as RRA began to emerge in the late 1970s, and were a result of the dissatisfaction with a number of elements of traditional field surveys.[2] First, there was a disillusion with questionnaire surveys which, to quote Chambers, 'tended to be long-drawn-out, tedious, a headache to administer, a nightmare to process and write up, inaccurate and unreliable in data obtained...' (Chambers, 1993:7). Second, there was the dissatisfaction with the usual biases; a spatial bias (preferably village surveys near cities), a personal bias (preferably men and local elites as respondents), a seasonal bias (preferably in the dry season), and a diplomatic bias (trying not to cause any offence). Third, there was the sometimes high cost of traditional survey methods. Apart from these three origins of RRA, there was the growing recognition by development professionals of the painfully obvious fact that 'simple' rural people were themselves knowledgeable on many subjects which touched their lives.

Although in the initial years such RRA surveys were proving very valuable, efficient and cost- and time-effective, most of those who used them were reluctant to write about it, fearing for their professional credibility. Consequently, they continued to use 'respectable' questionnaire surveys with proper numbers, purely to convince the establishment. However, as Chambers explains, by the mid-1980s, 'the RRA approaches and methods, when properly conducted, were more and more eliciting a range and quality of information and insights inaccessible through more traditional methods. Except when rushed and unself-critical, RRA came out better by criteria of cost-effectiveness, validity and reliability when it was compared with more conventional methods' (Chambers, 1992:8). In other words RRA, when well done, showed itself to be not just a second-best method, but a best overall.

RRA began and continues as a better way for outsiders (universities, aid agencies, NGOs) to learn, and to gain information and insight from rural people. It, thus, largely depends on local people's knowledge. Moreover, it is less exploitative than extractive questionnaire surveys. Later, by the late 1980s/early 1990s, Participatory Rural Appraisal (PRA) became to be established as a more developed but distinct form of RRA. PRA distinguishes itself from RRA in that it has the intention to enable local people to conduct their own analysis, and often to plan and take action themselves. Hence the shift in the mode of operation from extractive-elicitive in RRA to sharing-empowering in PRA, the shift in the role of the outsider from investigator to facilitator, the shift in the ownership, analysis and use of the information from outsiders to the local people themselves, and the shift from either individual or group activity in RRA to mainly group activity in PRA. Whereas in extractive research and RRA the outsiders (*we*) are still dominant, in PRA this is largely reversed and we encourage *them* to be dominant, and to largely determine the agenda. They are allowed and encouraged to use their knowledge in drawing maps, modelling, making diagrams, making estimates, etc... Both RRA and PRA, however, continue to have many overlapping features, but are being used in different contexts and with different objectives in view. In our field survey, RRA has been predominantly used.

So far as reliability is concerned, although every RRA (and PRA) practitioner must be familiar with the question 'But how does it compare with the *real* data?' (to rephrase the title of Gill's paper),[3] a large number of scholars and aid workers have been impressed by the correctness of such local knowledge and capabilities in RRA (and PRA). Wherever properly-done RRA or PRA research was followed by questionnaires to test the RRA/PRA results, the outcomes of both techniques have been impressively similar (see for example Franzel and Crawford 1987, Andy Inglis 1990, Gerard Gill 1992, Neela Mukherjee 1995, Ian Scoones 1995, PLA notes Oct. 1995). Moreover, although RRA (and PRA) research is much cheaper, much less time-consuming, excludes the possibility of badly designed or badly implemented questionnaires, does not require many people, does not give possible nightmares with data processing and analysis, and is much more informal, relaxing and far more fun for both interviewer and respondents, in many cases RRA research has not simply given more information, but also more diversified and more correct data than questionnaires have given. Any critique against RRA research data is usually centred around 'bias', with research results that are presumably being coloured by personal views and feelings. Patricia Norrish (among others) correctly questions whether there is any such thing as objectivity in research... in any research! (be it qualitative or quantitative) (Norrish, 1996).

However, examples of badly conducted and implemented RRA research, with rushed interviews and an insensitivity to the context or local culture, and to what is being said and how, have also occurred. It all depends on the individual practitioner. As Pretty, who identified a framework of criteria of RRA/PRA trustworthiness, observes, 'It will never be possible ... to be certain about the trustworthiness of criteria. The criteria themselves are value-bound, and so we cannot say that "x has a trustworthiness score of y points", but we can say that x is trustworthy because certain things happened during and after the investigation' (Pretty, 1995:1255-6).[4] Or as Mukherjee, comparing data collection through both PRA and questionnaires argues, that 'Any data collected is likely to have some errors. ... A weak methodology is one which is prone to more errors. It is important to identify the errors since they affect any estimation made from the data set and the conclusions and recommendations made on the basis of those estimates' (Mukherjee, 1995:39-40). For Mukherjee the major source of errors is bias, in its many forms.[5]

Looking back, application of the RRA technique in the villages of New Bombay has, I think, proved to be the right choice. First of all, as no similar study in the area had been done before, earlier information on family, social life style, employment and income, and more such things, was simply not available. Consequently, a questionnaire survey would possibly have started from a wrong set of questions, whereas now it has been possible to drastically restructure the set of questions in an early phase of the RRA survey. Second, as mentioned before, at the time of the fieldwork, CIDCO was preparing to conduct a large socio-economic survey in the villages based on questionnaires on its own. This was also a factor in the decision to do another kind of research, not based on questionnaires. Third, a questionnaire survey would have been far more expensive and time-consuming, and probably also less reliable, as a number of students or other helpers would have had to be engaged in taking the questionnaires, with possible mistakes, misunderstandings, or even neglect as a consequence. Fourth, in case of group discussions, several ideas and arguments could be written down, whereas with questionnaires the questionnaire form would have allowed only a single answer. Moreover, group discussions can give a very good idea of the things that live most strongly within the community. Fifth, the kind of data I was hoping to collect were mainly concerned with information on the lives of the local people, and the changes that had occurred for them personally and for their village habitat under influence of the development of the New Bombay area. Logically, as the main purpose of this part of my research was to extract information from the villagers, to learn from their experience and their feelings, RRA was preferred above PRA, although several typical PRA techniques have now and then also been used. Sixth, what has been experienced as most useful in this kind of research, and what has been felt as being a major advantage over questionnaires, is the fact that with semi-structured interviews the process of learning, analysing and theorising automatically starts from the very first talk, and that the information gathered from earlier interviews is systematically being used in subsequent interviews. The theoretical picture that gradually emerges on such a basis, is increasingly being enlarged and refined at the same time. With questionnaires, the first analytical results would probably be produced only from the moment that the entire fieldwork has already been fully completed.

Finally, but most importantly, the use of an informal and relaxed research technique such as RRA has proved to be much more pleasant and, partly because of that, much more extractive than questionnaires would ever have been. For a large number of villagers, I was the first outsider who had ever come to listen to their problems and feelings, the first outsider to whom they had the chance to express their possible anger and grievances. After the first reluctance they eagerly used that opportunity in a convincing manner. My role was limited to listening (real listening!) to their stories, to carefully bringing into the conversation those questions for which I was seeking an answer, and to making the villagers speak without holding back. In contrast to much of the earlier and more conventional survey research, it was my role to avoid any domination, any lecturing, and any interference. In contrast to questionnaire surveys for example, I could look into their eyes while listening to their stories, instead of constantly keeping my eyes on a questionnaire form. I could observe their non-verbal expressions, which often do say more than words. 'The fieldwork stick', which in earlier times used to be exclusively and firmly held by scholars, was partly being handed over to them. Hence the sometimes very long, very dynamic, and very animated conversations and discussions.

As is commonly argued, the major 'weakness' of qualitative research in general, and of RRA in this case in particular, is the fact that the kind of data of such surveys do not easily allow for nice tables full of figures, something which in some circles is still considered as a major condition of 'proper' field research. This is, of course, not the essential point. Nice-looking tables based on questionnaire data can be anything, from excellent to misleading and even absolutely wrong in terms of interpretation, as they may, for example, be the outcome of a wrong technique of data processing or incorrect use of statistical analysis.[6] Moreover, conventional questionnaire surveys mostly produce averages, not seldomly excluding any deviations which are considered abnormal. In RRA (and PRA), on the other hand, each view is important, although this does not obstruct the analysis in terms of an average picture. But, as we have mentioned before, data produced by application of RRA (and PRA) can, obviously, also be incorrectly gathered or seriously misinterpreted. In both cases, the quality of the survey and the interpretation of the data largely depends on the individual practitioner; in the case of a questionnaire survey primarily on technical capabilities (in the field) and technical and analytical capabilities (data processing and analysis) afterwards; in the case of RRA (and PRA) largely on personal capabilities, training, the ability to hand over the stick, the ability to adapt to a given socio-cultural context, and as Mason described it, on the ability to produce an analysis with rigour, with care, and with intellectual and strategic thinking.

Appendix 2: Tables produced from data SES-95/96

1. Type of houses
2. Yearwise occupation of tenements
3. Status of houses
4. Cost of houses under different categories
5. Size of dwelling units. 1. CIDCO constructed / 2. Others
6. Source of funds for financing cost of tenements
7. Average monthly instalment as % of household income
8. Average monthly rent paid for different size tenements
9. Last place of residence
10. Reasons for shifting to New Bombay
11. Households by different religion
12. Households by different mothertongue
13. Households by state of domicile
14. Number of CIDCO employees, PAP families, SC/ST/DT/NT/OBC families
15. Number of vehicles owned
16. Age and sex distribution of population
17. Level of education and age groups
18. Household size
19. Total earners in the family
20. Type of employment by income groups for working members, male/female
21. Household income by family size and comparison of per capita income
22. Distribution of population by occupation, male/female/combined
23. Location of workplaces by mode of transport
24. Destination and mode of transport for non-work trips
25. Purpose of non-work trips
26. Views of residetns on social facilities provided by CIDCO
27. Number of people having household business
28. Number of households having combined flats
29. Number of households having kept sub-tenant
30. Household income and size of flat (for CIDCO constructed and others)
31. Location of workplace and type of jobs
32. Level of education and medium of instruction

Appendix 3: Detailed household income figures BUDP Airoli (based on SES-95/96, report 21)

Income groups									
Sector	1	2	3	4	5	6	7	NM	Total
2	1	24	32	35	4	0	2	4	102
2a	3	23	29	11	3	0	0	4	73
2b	2	21	21	5	2	0	0	2	53
2c	2	12	17	4	2	0	0	1	38
2d	0	2	2	0	0	1	0	0	5
2e	2	10	7	1	1	1	0	0	22
3	1	3	3	8	2	0	0	0	17
3b	0	5	2	1	0	0	0	0	8
3j	1	8	3	4	1	0	0	0	17
4	10	53	188	115	11	4	1	5	387
4a	1	2	2	0	0	0	0	0	5
4b	1	9	9	3	2	0	0	2	26
4e	0	12	8	4	0	0	0	0	24
4f	0	11	10	8	1	0	0	0	30
4g	1	11	10	2	1	0	0	1	26
4m	1	15	8	5	2	0	0	1	32
4n	0	2	2	0	0	0	0	0	4
4p	0	6	2	1	0	0	0	0	9
4q	0	12	11	4	0	0	0	1	28
Total	26	241	366	211	32	6	3	21	906

Note: NM=not mentioned

Income groups: 1 = Rs. <1250 / 2 = Rs. 1251-2650 / 3 = Rs. 2651-4450 / 4 = Rs. 4451-7500 / 5 = Rs. 7501-10,000 / 6 = Rs. 10,001-15,000 / 7 = Rs. >15,001

Appendix 4: Water charges in New Bombay, and comparisons with adjacent regions

Metered water charges in New Bombay 1972-1996 (in Rs. per m³)

	Domestic use	Commercial use
1972-1978	Rs. 0.45	Rs. 1.50
1978-1980	Rs. 0.70	NA
1980- 1986 (a)	Rs. 1.50	NA
1986-1990 (a)	Rs. 2.10	NA
1990-1994 (a)	Rs. 2.80	Rs. 15.0
since Dec 1995	Rs. 3.65	Rs. 25.0

Source: MWSSB notifications 1994 and 1995; local engineering dept. CIDCO (Turbhe). Note: (a) exact period not certain; NA=not available

Fixed water charges in New Bombay 1996

Built-up area	Rate per month
upto 25 m²	Rs.37.00
25-40 m	Rs. 45.00
40-50 m²	Rs. 53.50
50-60 m²	Rs. 66.50
60-70 m²	Rs. 70.00
70-80 m²	Rs. 75.00
80 m² and above	Rs. 75.00

Source: CIDCO's revision of water rates (7/12/1995)

Water rates to various main categories (in Rs. per m³) in surrounding municipal corporations (as per 1/4/1994)

	Domestic use	Institutional use	Commercial use
New Bombay	*Rs. 2.80*	*Rs. 6.00*	*Rs. 15.00*
Greater Bombay	Rs. 0.60	Rs. 4.50	Rs. 12.00
Kalyan/Dombivali	Rs. 2.00	NA	Rs. 6.00
Thana	Rs. 1.30	Rs. 8.00	Rs. 15.00

Source: MWSSB. Note: NA=not available

Notes

¹ For a discussion on the possible advantages of qualitative interviewing over questionnaires see for example Mason, 1997:39-42. Mason's book is interesting because it provides a very accessible text based on a question-answer basis. All the major questions that one is confronted with at any stage in conducting qualitative fieldwork are being discussed and answered.

² Personally, I find the term *Rapid Rural Appraisal* a fairly unfortunate choice. RRA can, but does not have to be *rapid* (preferably, it should not be); it can, but does not have to be exclusively *rural*; and it can be, and is in many cases, much more than simply an *appraisal*. See for example Jo Abbot's comments on this issue.

³ Curiously, this question is always being asked to RRA or PRA researchers, and rarely or never to those who follow conventional methods of quantitative research.

⁴ Some of the PRA criteria Pretty identified also count for good RRA research, for example a prolonged and/or intense engagement between the various actors, a persistent and critical observation, an analysis and expression of difference, and triangulation (for cross-checking information through, for example, the use of secondary data).

⁵ Some of the sources of error giving rise to bias that Mukherjee puts forward are the expectations of the researcher, his or her age, race, sex, profession, religion, dress, etc., the selection of area, place, time, and people, and the wording or even the position of the questions.

⁶ The Essex Summer School in Statistical Analysis, to which I was a participant in 1994, has been enough proof to realise that the way in which quantitative data are being collected, processed and analysed can be absolutely wrong in at least a hundred different ways. Campbel, Shrestha and Stone, among others, have provided some proof of badly conducted quantitative research, i.c. in Nepal (in that case apparently due to linguistic difficulties with the questionnaire vocabulary, the questions were simply not properly understood).

Printed and bound by CPI Group (UK) Ltd, Croydon, CR0 4YY

21/10/2024

01777087-0012

Notes

[1] For a discussion on the possible advantages of qualitative interviewing over questionnaires see for example Mason, 1997:39-42. Mason's book is interesting because it provides a very accessible text based on a question-answer basis. All the major questions that one is confronted with at any stage in conducting qualitative fieldwork are being discussed and answered.

[2] Personally, I find the term *Rapid Rural Appraisal* a fairly unfortunate choice. RRA can, but does not have to be *rapid* (preferably, it should not be); it can, but does not have to be exclusively *rural*; and it can be, and is in many cases, much more than simply an *appraisal*. See for example Jo Abbot's comments on this issue.

[3] Curiously, this question is always being asked to RRA or PRA researchers, and rarely or never to those who follow conventional methods of quantitative research.

[4] Some of the PRA criteria Pretty identified also count for good RRA research, for example a prolonged and/or intense engagement between the various actors, a persistent and critical observation, an analysis and expression of difference, and triangulation (for cross-checking information through, for example, the use of secondary data).

[5] Some of the sources of error giving rise to bias that Mukherjee puts forward are the expectations of the researcher, his or her age, race, sex, profession, religion, dress, etc., the selection of area, place, time, and people, and the wording or even the position of the questions.

[6] The Essex Summer School in Statistical Analysis, to which I was a participant in 1994, has been enough proof to realise that the way in which quantitative data are being collected, processed and analysed can be absolutely wrong in at least a hundred different ways. Campbel, Shrestha and Stone, among others, have provided some proof of badly conducted quantitative research, i.c. in Nepal (in that case apparently due to linguistic difficulties with the questionnaire vocabulary, the questions were simply not properly understood).

Printed and bound by CPI Group (UK) Ltd, Croydon, CR0 4YY

21/10/2024

01777087-0012